AF589376

Fish Population Dynamics and Stock Assessment

THE AUTHORS

Dr N Jayakumar started his career as Assistant Professor in 2005 and he is presently working as Assistant Professor and Head in the Department of Fisheries Resource Management in Tamil Nadu Dr. M.G.R. Fisheries College and Research Institute, Ponneri, a constituent of Tamil Nadu Dr. J. Jayalalithaa Fisheries University. Earlier, he served as Fisheries Science Teacher under the Ministry of Education in the Republic of Maldives for about six years from 1999 to 2005. In addition, he secured an international fellowship for undergoing a PG course in Tropical Coastal Ecology, Management and Conservation in Kenya under Kenya-Belgium Project in Marine Sciences during 1998. He served as Principal Investigator in a NADP scheme and Co-Principal Investigator in four different research schemes. In total, he has gained a teaching experience of 19 years and research experience of 13 years. He has been teaching various courses on fisheries biology and resource management to the students of B.F.Sc., M.F.Sc and Ph.D (Fisheries Resource Management). He prepared three ICAR e-courses for the students of B.F.Sc. He published about 40 research papers in peer-reviewed journals and 12 teaching / practical / training manuals. He is the recipient of "Best Teacher Award" from Tamil Nadu Fisheries University (2015) and "Outstanding Faculty Award" from Venus International Foundation (2017).

Mr R Durairaja is presently working as Assistant Professor in the Department of Fisheries Resource Management, Tamil Nadu Dr. M.G.R. Fisheries College and Research Institute, Tamil Nadu Dr. J. Jayalalithaa Fisheries University, Ponneri, Thiruvallr District. He has been involving in teaching, research and extension activities relevant to fisheries resource management for the last three years. He did his post-graduation (M.F.Sc) in Fisheries Biology and Resource Management in West Bengal University of Animal and Fishery Sciences, Kolkata during 2003 – 2005. He is involved in research on the biodiversity and stock assessment of fishes in Pulicat Lake. He serves as a life time member in IFSI, CIFRI, Barrackpur and ASTS, New Delhi and member of WAS. He had been to Canada for the purpose of attending "The Dialogue on International Food Security" held at the University of Alberta in Edmonton in 2014.

Dr P Jawahar is Professor of Fisheries Resource Management at Tamil Nadu Dr. J. Jayalalithaa Fisheries University, India. He completed his Ph.D. degree in Fisheries Biology and Capture Fisheries. His main areas of interest are marine biodiversity, stock assessment and co-management of fisheries resources. He is having twenty years of teaching experience in Fisheries Biology and Resource Management and currently working on marine resources of Gulf of Mannar. He has published eighty research articles, twenty manuals, two books, five booklets and many popular articles. He has also carried out several research projects. Being member in Research Advisory Committee, GoMBRT, he has contributed actively to the the management of marine living resources of Gulf of Mannar. He is also actively participating in co-management of lobster resources and fisheries resources.

Dr S Felix, Ph.D., Vice Chancellor, Tamil Nadu Dr. J. Jayalalithaa Fisheries University, Nagapattinam is having experience of more than 35 years in fisheries and aquaculture in teaching, research, extension and administration. He has organized more than 20 Nos. of seminar/conference/symposium at National and International levels. Currently he is also the President (2018-19) of World Aquaculture Society, Asian Pacific Chapter and Chairman, ICAR's BSMA Committee (2018-19). He has more than 60 research papers published in National and International journals. He has authored more than 10 books and 25 manuals. He is specialized in the area of advanced systems in aquaculture and aquariculture such as raceway, biofloc technology, RAS, etc,. He has operated around 20 externally funded research projects.

Fish Population Dynamics and Stock Assessment

by

N Jayakumar

R Durairaja

P Jawahar

S Felix

Tamil Nadu Dr. M.G.R. Fisheries College & Research Institute
Tamil Nadu Dr. J. Jayalalithaa Fisheries University
Ponneri, Tiruvallur District, Tamil Nadu

DAYA PUBLISHING HOUSE®
A Division of
ASTRAL INTERNATIONAL PVT. LTD.
New Delhi – 110 002

ISBN: 9789388173254 (Int. Edition)

Published by : **Daya Publishing House®**
A Division of
Astral International Pvt. Ltd.
– ISO 9001:2015 Certified Company –
4736/23, Ansari Road, Darya Ganj
New Delhi-110 002
Ph. 011-43549197, 23278134
E-mail: info@astralint.com
Website: www.astralint.com

Laser Typesetting : **Classic Computer Services,** Delhi - 110 035

Printed at : **Neelam Graphics, Delhi - 110007**

डॉ. जे. के. जेना
उप महानिदेशक (मत्स्य विज्ञान)

Dr. J. K. Jena
Deputy Director General (Fisheries Science)

भारतीय कृषि अनुसंधान परिषद
कृषि अनुसंधान भवन-II, पूसा, नई दिल्ली 110 012
INDIAN COUNCIL OF AGRICULTURAL RESEARCH
KRISHI ANUSANDHAN BHAVAN-II, PUSA, NEW DELHI - 110 012

Ph. : 91-11-25846738 (O), Fax : 91-11-25841955
E-mail: ddgfs.icar@gov.in

Foreword

Population is usually defined as the aggregates of individuals of the same species, living in a well-defined geographical region. Such population gets cut off from another distinct population of the same species through a discontinuity between their distributional ranges. Fish population dynamics involves the quantitative change in a population, which is controlled by recruitment, natural mortality by predation, disease & old age, fishing mortality and migration. Such studies provide understanding on changing fishery patterns and associated issues viz., habitat destruction, predation, overfishing etc. which largely affect the fisheries. The subject has been instrumental in determining sustainable fish yields by our fisheries researchers, which ultimately help in devising changes for effective management of fisheries in our open-waters.

The stock assessment, which uses various statistical and mathematical calculations to make quantitative predictions about the fish populations, aims to provide advice on the optimum exploitation of the fisheries resources. It gives an understanding about the abundance of a stock and also level of exploitation, thereby suggesting appropriate management measures. It is necessary that the fish stock assessment is undertaken for each stock separately and the results may be pooled subsequently for the assessment of a multispecies fishery.

I sincerely believe that the present book on '***Fish Population Dynamics and Stock Assessment***' would greatly help the fisheries students and scientists in acquiring adequate knowledge on population dynamics and in solving the problems of fish stock assessment.

I congratulate and compliment the authors for their sincere efforts in bringing out this useful publication.

(J.K. Jena)

Message

India is blessed with vast resources in terms of 8,129 km long coast line, 0.53 million km^2 of continental shelf and 2.02 million km^2 of exclusive economic zone, the marine fisheries in India has been playing a pivotal role in meeting the demands of fish over the years. The marine fisheries sector in India has registered a phenomenal growth during the last five decades both quantitatively and qualitatively.

Fish stock assessment studies have made remarkable progress in the last few decades. Accurate assessment of fish stock will help in optimum exploitation. As most of our commercially important marine fishes have started showing signs of depletion due to overexploitation, the assessment of fish stock is urgently needed to find out sustainable level of existing fish stock, to assess the standing stock biomass of fish in our seas and for the prediction of future fish yields. Mathematical models are now-a-days extensively used in assessing the fish stock.

A comprehensive book dealing with methods in fish population dynamics and stock assessment is urgently needed. This book which has been prepared with a number of examples relating to tropical fish stocks, helps fishery students and scientists in data analysis and in providing necessary skills in solving fish stock assessment problems and in my view, it will be of immense use for researchers and students of Universities and colleges where fisheries biology courses are taught. Glossaries on the most commonly used technical terminologies are provided to facilitate easy learning for the beginners.

S. Felix
Vice Chancellor
Tamil Nadu Dr. J. Jayalalithaa Fisheries University

Preface

Fish stock assessment in tropical waters has made remarkable renaissance in the last few decades. As most of the stocks of commercially important species of India are getting reduced due to increase in fishing effort, cut down in effort is needed. Further, restriction in the catch of juvenile fish could be advised on the basis of knowledge about the growth rate during the first years of the life of the fish. In tropical countries like ours, where we have multi-species fisheries, estimate of age and growth, mortality, standing stock biomass and yield per recruit are more difficult because of the prolific breeding nature of the species. Hence methods normally used for the stock assessment for temperate fishes are not applicable to tropical species. Keeping the above in view, a book entitled, "Fish Population Dynamics and Stock Assessment" has been prepared. This publication brings out all the methods in detail and emphasis has been placed on these methods which are tailored to the need of students and researchers and scientists involved in Fish Stock Assessment.

We are greatly indebted to Dr. S. Felix, Vice Chancellor, Tamil Nadu Dr. J. Jayalalithaa Fisheries University, Nagapattinam for according permission for publishing this book. We sincerely thank Dr. B. Ahilan, Dean i/c., TN Dr. MGR FC&RI, Ponneri for his encouragement.

We would welcome and very much appreciate constructive criticism and comments for the improvement of this book in the future edition.

Authors

Contents

Symbols

a: Multiplicative factor generally used in exponential relationships linking L and W ($W=aL^b$); Y intercept of linear regression

a: Swept area (effective path swept by trawl)

a': intercept of functional regression

b': Slope of functional regression

b: Exponent of a length-weight relationships; slope of linear regression

A: Annual mortality rate

B: Biomass, or stock size in weight of a population fish

Bo: Biomass, prior to any fishing; unfished biomass

B/R: Biomass per recruit

C: Catch in numbers; parameter expressing the amplitude of seasonal growth oscillations in the VBGF; a constant

C/F: Catch per unit of effort (also: CPUF)

e: Base of the natural (or Naperian) logarithms; e=2.71828

E: Exploitation rate; E=F/Z

E0.1: Level of exploitation at which the marginal increases in yield per recruit reaches 1/10th of the marginal increases computed at a very low value of E.

E0.5: Exploitation level which results in a reduction of the unexploited biomass by 50 per cent.

Emax: Exploitation level which maximizes Y/R or Y'/R

f: Fishing effort

F: Instantaneous rate of fishing mortality, *i.e.* rate of morality, of dimension time^{-1}, due to fishing gears, whether or not the killed fish are caught and landed.

Fmax: Fishing mortality generating maximum yield per recruit

FMSY: Fishing effort generating MSY

H: Natural mortality factor in Jone's length based cohort analysis

K: Curvature parameter of the VBGF; rate of dimension time-1 at which L is approached

L: Length increment; width of length class in grouped data

In: log_e - logarithm of base e; Naperian logarithm

log: Logarithm with base 10

$L_{(t)}$: Length of the fish at age t in years

$L_{(t)}$: Mean growth length

$L_{(t+1)}$: Length of the fish in the subsequent years

L1-L2: Length class

L1-L2: From length L1 to length L2

L: Parameter of the VBGF< expressing the asymptotic length, *i.e.* the mean length the fish in a population or stock would reach if they were to grow indefinitely; often close to the length of a very old fish

L∞: Asymptotic length, Length infinity

L_{max}: The largest fish

L/F: Length-frequency or length-frequency samples

L25: Length at which25 percent of the fish will be vulnerable to the gear (left-hand selection)

L50: Length at which 50 percent of the fish will be vulnerable to the gear (left-hand selection)

L75: Length at which 75 percent of the fish will be vulnerable to the gear (left-hand selection)

L': Some length for which all fish of that length and large are under full exploitation.

Lc: Mean length of fish or entire catch

Lm: Length at first maturity (or "massive maturation")

Lmass: Upper class limit of the largest class in a sample; maximum length reached by the fish of a given stock; may also be predicated from the largest specimens of several samples using the extreme value theorem

Lmin: Smallest length represented in one or several sample

m: K/Z

M: Instantaneous rate of natural mortality which occur due to natural cause and not due to fishing

MEY: Maximum Economic Yield

ML: "Midlength" or length class midpoint; midpoint of a class interval

MPA: Modal progression analysis

MSY: Maximum Sustainable Yield, a frequently criticized but still useful concept

MSE: Maximum Sustainable Economic Yield

N: Number of fish in a length class, Number of survivors (VPA)

N_o: Initial number of individuals in a population exposed to some source of a mortality

N_r: Initial population (in number) of a cohort, *i.e.* at age t_1

N_t: Number of fish in the oldest age group of a cohort or population ("terminal population")

N(t): Number of survivors of cohort attaining age

N(Tr): Number of recruits to the fishery.

i': (phi-prime), In K+2* In L∞

q: Condition factor, constant in length – weight relationship

q: Catchability coefficient, Proportionality coefficient

R: Recruitment; also true relative abundance of an age group; number of recruits (NTR)

R_{25}: Length at which 25 percent of the fish will no longer be vulnerable to the gear (right-hand selection, or deselection)

R_{50}: Length at which 50 percent of the fish will no longer be vulnerable to the gear (right-hand selection, or deselection)

R_{75}: Length at which 75 percent of the fish will no longer be vulnerable to the gear (right-hand selection, or deselection)

S: Annual survival rate

SF: Selection factor

t: A given time or age (normally expressed in years); absolute age of a fish, for example an estimate from daily otolith rings; age corresponding to L_t

t': Some age for which all fish of that age and older are under full exploitation

ti: Mean age at length i

t_m: Age at marking, corresponding to t_m

t_{mass}: Mean age to which most fish reach maturity (massive naturation at 50 per cent level)

t_{max}: Longevity (in the wild); also oldest (calculated) age in a sample

t_{min}: Youngest calculated age in a sample

t_o: Parameter of the VBGF, expressing the theoretical age at which the fish would have a length zero if they had always grown according to that equation ('t_o'almost always takes non-zero, negative values, because small fish usually grow faster than predicted by the VBGF)

t_s: Parameter of the seasonally oscillating version of the VBGF (Summer point 0.1)

t_w: Parameter of the seasonally oscillating version of the VBGF (Winter point 0.1)

T: Mean ambient temperature, in ^{0}C.

TAC: Total Allowable Catch

Tc: Age at first capture (start of explaited phase)

TL: Total length

VBGF: von Bertalanffy Growth Function

VPA: Virtual population analysis

W_i: Mean weight of fish within a given length class i

W: Live weight of an individual fish

W∞: Asymptotic weight, weight infinity

W_{max}: Maximum weight reached by the fish of a given stock

WP: 'Winter point' in the seasonalized VBGF, the time of the year when growth rate is slowest; equivalent to t_s+0.5 year

x: Independent variable

y: Dependent variable

Y/R: Yield per recruit (Beverton and Holt)

(Y/R): Relative yield per recruit (Beverton and Holt)

Y/R_{max}: Maximum yield per recruit achievable under a given fishing regime

Z: Instantaneous rate of total mortality

Δ: Denotes a difference between two lengths, time *etc.*

Δt: Length increment

Δ1/Δt: Growth rate expressed as difference

Δ t: Time difference, such as time needed by average fish to grow from lower to upper limit of length class.

Chapter 1

Introduction

Population is usually defined as aggregates of individuals of the same species, inhabiting a well-defined geographical area. Such population gets cut off from another discrete population of the same species through a discontinuity between their distributional ranges. A population can be considered a unit stock if its members exhibit sufficiently uniform characteristics in respect of their growth, spawning, natural mortality, response to fishing, morphometry, meristic counts, serological properties *etc.*

Mechanics is the science of statics and dynamics of bodies. Dynamics is concerned with the geometrical aspects of motions of bodies without reference to what causes them (Kinematics) and also the effect of forces as they condition these motions (Kinetics). On the other hand, statics deals with bodies which are at rest and with forces in equilibrium. The concept of dynamics applies to any branch of science in which forces are considered. With an appropriate prefix, the term 'dynamics' is expressed as hydrodynamics, aerodynamics, thermodynamics, population dynamics, *etc.* indicating the particular branch of science to which its laws generally apply.

Population Dynamics

Population dynamics is a branch of life science which is concerned with the short and long-term changes in the size and age composition of populations, and the biological and environmental processes influencing those changes. It describes the way in which a given population grows and shrinks over time as controlled by birth, death and emigration and immigration. It also studies aging of population and population decline.

Population dynamics has traditionally been the dominant branch of mathematical biology, which has a history of more than 210 years, although more recently the scope of mathematical biology has greatly expanded. The first principle of population dynamics is widely regarded as the exponential law of Malthus, as modeled by the Malthusian growth model.

A more general model formulation was proposed by F.J. Richards in 1959, further expanded by Simon Hopkins, in which the models of Gompertz, Verhulst and also Ludwig von Bertalanffy are covered as special cases of the general formulation. The Lotka-Volterra predator-prey equations are another famous example. The best mathematical model that governs the population dynamics of any given species is called the exponential model. In fisheries and wildlife management, population is affected by three dynamic rate functions, such as natality or birth rate, population growth rate and mortality rate.

Stock Assessment

Stock assessment is an important tool in fisheries management. In particular, to ensure contained healthy fish stocks, measurements of the spawning stock biomass (the stock population capable of reproducing) allows sensible conservation strategies to be developed and maintained through the application of sustainable fishing quotas.

In fisheries management, stock refers to a harvested or managed unit of a fish. Typically stocks are divided based on geographical location and not based on individual population. Stocks are not always composed of a single species. Stocks can be composed of multiple species due to their being harvested together or as a form of convenience for managers. An example of a multi-species stock is river herring. Anchovies and blueback herring are labeled as river herring for management purposes due to their similar physical appearances and being harvested together. Individuals within a stock are subdivided into cohorts. A cohort is a group of fish born in the same year within a population or stock.

Concept of Population and Unit Stock

Population is the breeding unit of species having common spawning grounds. A population is a group of individuals of the same species that live together in the same place, and that possess an average set of properties, such as birth rates and death rates. This definition recognizes that populations are made up of individual organisms but does not require that we know which individuals give birth or die, or where they are located in space. Instead the population is characterized by average birth and death rates, and variability in these averages is treated as a statistical property of the population. Most definitions of population have some kind of spatial reference. The simplest and least restrictive of these is that a population is a group of individuals of the same species that live together in a particular area (*e.g.,* Roughgarden, 1989). However, even though this definition is widely used by ecologists, it gives rise to serious difficulties and misinterpretations. A more rigorous definition should define the spatial dimension more precisely; for example, a group of individuals of the same species that live together in an area of sufficient size that all the requirements for reproduction, survival and migration can be met (*e.g.*, Huffaker *et al.*, 1984). It may be helpful to conceive an idea of sufficient size such that the rates of emigration out of the area and immigration into the area are roughly balanced. Whatever method is used to define the appropriate size of an area within which the population of a particular organism exists, it is important that most of the change in population size or density is due to births and deaths rather than immigration and emigration because the theory of population dynamics is based on this assumption.

Local population is a group of organisms of the same species that live together in an area where there is a high probability of interbreeding. Also called a sub-population or, in systematic, a deme (populations of related species). Emigration and immigration from local populations need not necessarily be balanced and there may be fairly high probability of local extinction.

- ✰ Meta-population is a group of populations that share occasional migrants
- ✰ Absolute population is an estimate of the total number of organisms in an area
- ✰ Population density is an estimate of the number of organisms per unit area (*e.g.*, hectare) or unit of habitat (*e.g.*, kilogram of soil)
- ✰ Relative population is an estimate of the number of organisms caught in nets or traps but which cannot be related to area in any way

Size of population is determined by growth which is influenced by natality ('+' influence), mortality ('-' influence) and dispersion ('+ & -' influence). Various ecological density dependent and density independent factors are the factors that control the growth. For the purpose of fitting yield models, males and females of the populations are treated as separate units, if differences exist in all morphological and biological characters. For modelling purposes, population of different species coexisting in a particular area having similar characters can sometimes be treated as a single stock. Assessment could be made on stock of a species, only when biology of species is clearly understood which includes its feeding, growth, and spawning and migration habits.

Concept of Stock

Unit stock is a self-contained and self-perpetuating group, having defined geographical limits of spawning and gene exchange within stocks having same growth patterns and mortality rates. They should have same morphometry - meristics and spawning patterns and season.

In terms of stock assessment, stock is designated as a subset of species. For fishery management purposes, stock is a sub group of species. Stock includes portion of a population. Fish stocks are sub-populations of a particular species of fish, for which intrinsic parameters (growth, recruitment, mortality and fishing mortality) are the only significant factors in determining the stocks, while extrinsic factors (immigration and emigration) are considered to be insignificant. All species have geographic limits to their distribution, which are determined by their tolerance to environmental conditions, and their ability to compete successfully with other species. In marine environments, this may be less evident than on land because there are fewer topographical boundaries, however, discontinuities still exist, produced for example by meso-scale and sub-mesoscale circulations that minimize long-distance dispersal of fish larvae.

For fishes, it is rare for an individual to reproduce randomly with all other individuals of that species within its biological range. There is a tendency to form a structured series of discrete populations, which have a degree of reproductive isolation from each other in space, in time, or in both. This isolation is reflected in the development between sub-populations of genetic differences, morphological variations and exposure to different chemical regimes and parasitic species. Sub-populations also respond to fishing in such a way that fishing on one population appears to have no effect on the population dynamics of a neighbouring population.

The currently accepted definition of a stock in fisheries science is that of (Begg *et al.,* 1999), "... [A "stock"] describes characteristics of semi-discrete groups of fish with some definable attributes which are of interest to fishery managers". Stock identification is a field of fisheries science, which aims to identify these subpopulations, based on a number of techniques.

Straddling Stock

The United Nations defines straddling stocks as "stocks of fish such as Pollock, which migrate between, or occur in both, the Exclusive Economic Zone (EEZ) of one or more states and the high seas". Sovereign responsibility must be worked out in collaboration with neighboring coastal states and fishing entities.

Straddling stock can be compared with transboundary stock. Straddling stock range both within an EEZ as well as in the high seas. Transboundary stock range in the EEZs of at least two countries. A stock can be both transboundary and straddling. These stocks are usually pelagic, rather than demersal. Pelagic species are more mobile and their movements are influenced by ocean current, temperature and the availability of food.

Characteristics of Mixed Stock

A mixed stock fishery is a fishery whose stock consists of fish that are of a variety of ages, sizes, species, geographic or genetic origins or any combination of these variables. Mixed stock fisheries offer a challenge to fisheries managers due to the difficulty in targeting fish of a specific type using many commercial fishing methods. Mixed stock of a species has significant differences in morphological and biological characteristics. The mixed stock population may unite into a unit stock by gene flow.

The morphometric characters and meristic characters are to be taken for different stocks/populations of a species. Using Principal Component Analysis (PCA) statistical test, the population are to be segregated. Mixed stock of a species is common in tropical waters. Assessment of such tropical stocks becomes difficult by using conventional models. Refinements are needed for proper assessment. The aging becomes difficult in a mixed stock of tropical species. Usually many cohorts are released from a tropical stock; hence the assessment becomes more problematical on tropical stocks of a mixed species. Mixed stock of a species has varied spawning seasons and the calculation of length at minimum maturity and growth gets varied.

Limitations in the Assessment of Tropical Fish Stocks

For calculation of stock assessment of mixed stock of a species, the conventional models framed for temperate species should be used in caution. The mortality parameters such as Z, F and M, fishing effort and growth parameters are to be calculated separately, as these parameters form input data for estimation of stock in conventional models.

Chapter 2

Principles of Stock Assessment

The overall aim of fisheries science is to provide information to managers on the state and life history of the stocks. This information feeds into the decision making process. Fisheries science, economic, social and political considerations all have an impact on the final management decision. Stock assessment involves using mathematical and statistical models to examine the retrospective development of the stock and to make quantitative predictions to address the following fisheries management questions:

1. What is the current state of the stock?
2. What has happened to the stock in the past?
3 What will happen to the stock in the future under alternative management choices?

Stock Assessment

It is the process of collecting and analyzing demographic information about fish populations to describe the conditions or status of a fish stock. The result of a stock assessment is a report that often includes an estimation of the amount or abundance of the resource, an estimation of the rate at which it is being removed due to harvesting and other causes, and one or more reference levels of harvesting rate and/or abundance at which the stock can maintain itself in the long term. Stock assessment often contain short-term (1-5 years, typically) projections or prognoses for the stock under a number of different scenarios. This information on resource status is used by managers to determine what actions are needed to promote the best use of our living marine resources.

Stock assessment reports describe a range of life history characteristics for a given species, including age, growth, natural mortality, sexual maturity and reproduction, stock boundaries, diet preferences, habitat characteristics, species interactions, and environmental factors that may affect the species. Assessment reports also include descriptions of the fishery for a species, using information from both scientists and fishermen. Additionally, stock assessments describe the assessment model, or the collection of mathematical and statistical techniques that were used to perform the stock assessment.

Stock assessment analyses rely on various sources of information to estimate resource abundance and population trends. The principal information comes from the commercial and recreational fisheries (fishery-dependent information). For example, the quantity of fish caught and the individual sizes of the fish, their biological characteristics (*e.g.* age, maturity, and sex), and the ratio of fish caught to the time spent for fishing (catch per unit of effort) are basic data for stock assessments. Understanding the natural history of the harvested species and the other species with which they interact is crucial to understanding the population dynamics of living marine resources.

Fish Abundance

Fish abundance or population size can be expressed as either the number of fish or the total fish weight (or biomass). Increases in the amount of fish are determined by body growth of individual fish in the population, and the addition or recruitment of new generations of young fish (*i.e.* recruits; recruits from the same year are said to comprise a year-class [or cohort]). Those gains must then be balanced against the proportion of the population removed by harvesting (called fishing mortality, F) and other losses due, for example, to predation, starvation, or disease (called natural mortality, M). In stock assessment work, removals of fish from the population are commonly expressed in terms of rates within a time period. The fishing mortality rate is a function of fishing effort, which includes the amount, type, and effectiveness of fishing gear and the time spent for fishing. Catch per unit of effort (CPUE) is an index showing the ratio of a catch of fish, in numbers or in weight, and a standard measure of the fishing effort expended to catch them.

Surplus Production

Surplus production (or production) is the total weight of fish that can be removed by fishing without changing the size of the population. It is calculated as the sum of the growth in weight of individuals in a population, plus the addition of biomass from new recruits, minus the biomass of animals lost to natural mortality.

Production Rate

The **production rate** is expressed as a proportion of the population size or biomass. The production rate can be highly variable owing to environmental fluctuations, predation, and other biological interactions with other populations.

On average, production decreases at low and high population sizes, and biomass decreases as the amount of fishing effort increases. This means there is a relationship between average production and fishing effort.

Production Functions

The relationship between average production and fishing effort is known as the production function. Production functions are the basis for certain important concepts like: maximum sustainable yield. In addition, the term stock level is employed as a biological reference for determining resource status relative to the biomass that would on average support the sustainable yield. Recent average yield also is reported in order to allow comparison of the current situation to the sustainable yield.

Many other reference levels are used as benchmarks for guiding management decisions. A number of these are expressed as fishing mortality rate levels that would achieve specific results from the average recruit to the fishery if the stock were subjected to fishing at those rates indefinitely. Some of these benchmarks are used to index potential fishery production, and others are used to index potential reproductive output. F_{max} is the fishing mortality rate that maximizes the yield obtained from the average recruit. Growth overfishing occurs over the range of fishing mortality, at which the losses in weight from total mortality exceed the gain in weight due to growth. This range is defined as beyond F_{max}.

Maximum Sustainable Yield (MSY)

MSY is the maximum long-term average yield that can be achieved through conscientious stewardship by controlling F through regulating fishing effort or total catch levels. MSY is a reference point for judging the potential of the resource.

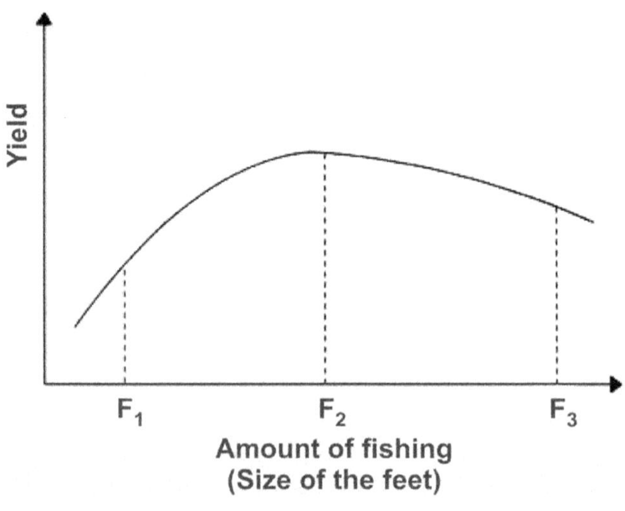

Figure 1

However, it is not necessarily the goal of fishery managers to always set the maximum yield. Other factors influence the choice of a management objective, such as socio-economic considerations or conservation and ecosystem concerns for other marine life indirectly affected by fishery harvests. One of the major objectives of fish stock assessment is regularisation or optimisation of effort.

Standardization of fishing effort – increasing the effort in the long term gives highest yield. The basic principle is smaller a population, when effort is not put at required level, the greater will be production. At optimal level of effort, the production will be maximum for a given Fmsy.

For all practical purposes, the effort should be monitored at various levels of fishing mortality. The basic principle in stock assessment models is to provide estimates of optimum yield. The environmental factors, economic factors are to be taken into account for arriving at appropriate management decisions. When intensity of fishing is not monitored, the population of a stock show sign of depletion and there will be a decrease in mean length of fish and length at minimum maturity.

In principle as age increases the number of survivors will be less.

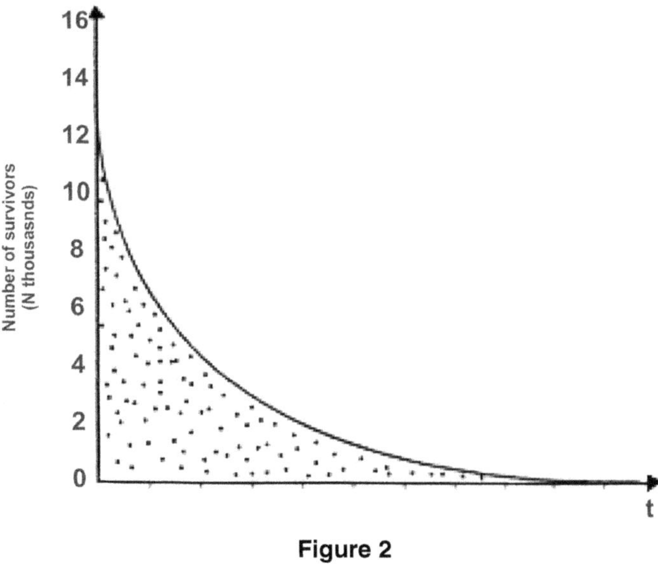

Figure 2

In a virgin (unfished) biomass, the population is at equilibrium. The population neither grows nor declines which means that each year's recruitment is balanced by each year losses due to mortality.

When a stock is brought below B∞ level, the population grows. The maximum surplus production is achieved when the stock biomass is reduced to half of the level of B∞.

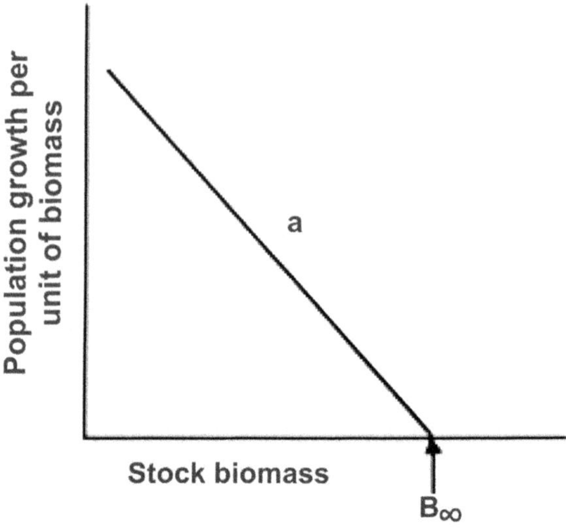

Figure 3

In principle, the highest yield achieved on a long term basis for a particular effort is the Fmsy and the corresponding yield is MSY or Maximum Sustainable Yield.

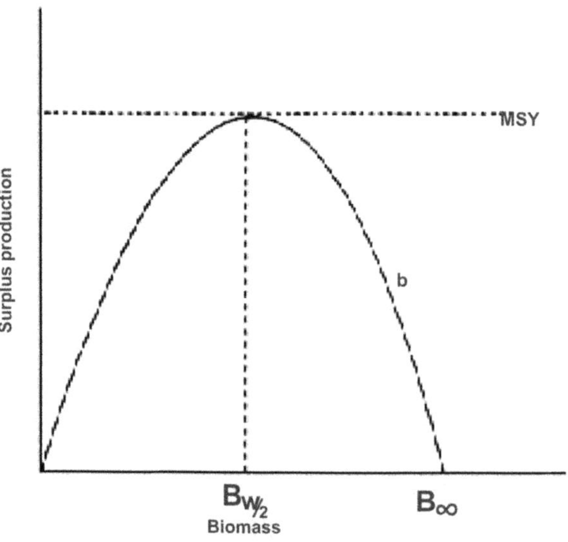

Figure 4

Chapter 3

Data Requirement of Stock Assessment

The data used in stock assessments can be classified as fishery-dependent data or fishery-independent data.

Fishery-Dependent Data

Fishery-dependent data is collected from the fishery itself, using both commercial and recreational sources. There are a variety of methods for obtaining fishery-dependent data. The most common approach is to use recorded landings. Landings are a record of the amount of fish sold and the numbers are typically reported in total weight. Another common mode for acquiring fishery-dependent data is through portside sampling of the catch of both recreational and commercial fishermen to obtain age and length information on the stock. Other less common methods for obtaining data is through the use of onboard observers, self-reporting, telephone surveys, and vessel-monitoring surveys.

Fishery-Independent Data

Fishery-independent data is obtained in the absence of any fishing activity. The majority of this data is collected by state and federal agencies. A wide variety of methods and gear types are used to acquire fishery-independent data. Sampling equipment can include trawls, seines, acoustic and/or video surveys. The study may focus on a single species, multiple species, or a specific age range or cohort. Regardless of the method or approach, these surveys provide managers with an estimate of abundance. Mark and recapture studies are commonly used to estimate movement, migration, and growth rate, natural mortality, and discard mortality. Stock assessments are often completed using both fishery-dependent and fishery-independent data. Stock assessments provide fisheries managers with the information that is used in the regulation of a fish stock. Biological and fisheries data are collected in a stock assessment.

Biological Data

The biological data include details on the age structure of the stock, age at first spawning, fecundity, ratio of males to females in the stock, natural mortality (M), fishing mortality (F), and growth rate of the fish, spawning behaviour, critical habitats, migratory habits, food preferences, and an estimate of either the total population or total biomass of the stock.

Fisheries Data

The fisheries data include the kinds of fishermen in the fishery, commercial versus recreational, and the gear used, amount of fish caught by fishing, age structure of the fish, sex ratio, how the fish are marketed, the value of the fish to the different fishermen groups, and the time and geographic location of the best catches. Further, geographical boundaries of different stocks or populations are also defined in the assessment. From the combined biological and fisheries data, the current status and condition of the stock is defined and managers use this assessment to predict how in the future, stocks will respond to varying levels of fishing pressure. Ultimately, managers want to reduce the levels of overfishing that occurs and restore stocks that have been overfished. For fish stock assessment studies, the growth of the fish and its age form the most necessary primary data. The age and growth data of a stock goes as input parameters for the estimation of growth parameters and mortality parameters. These two parameters along with selection parameters goes as input data for prediction models which will be dealt in the coming chapters. Fish stock assessment should be made sex-wise/stock-wise separately. The results should be subsequently pooled for optimization of assessment.

Basic Elements for the Description of a Fishery

1. The input data, namely fishing effort. (Fishing effort is the product of the amount of gear in use times the duration of fishing activity.)

   Examples are – Trawling hours, trap days, diving hours, hook hours *etc.*
2. The output data *i.e.* fish landed in fish landing centres and
3. Process which link the input and output data. Usually processes consist of biological process and the fishing operations.
4. Models are processes that connect input and output data. Prediction could be made possible by processing input data by using models.

In general stock assessment models are grouped into two:

1. Analytical model
2. Holistic model (*e.g.* surplus production model, box model, swept area method)

For Analytical Model

- ✰ Input – Age data (by hard parts, length frequency data (LFD or tagging).
- ✰ Process – Estimation of age and growth using the hard parts, LFD data.
- ✰ Growth parameters
- ✰ Mortality parameters
- ✰ Selection parameters
- ✰ Process – Yield estimation by different values of fishing mortality for a range of exploitation levels.
- ✰ Output - Estimation of MSY, standing stock biomass. Relative yield per recruit and to ascertain optimum mesh size of net for a given stock in question

For Surplus Production Model

- ✰ Input: Catch and effort.
- ✰ Process: Estimation of CPUE for different years.
- ✰ Output: MSY, FMSY

For Swept Area Method

- ✰ Input: Survey data in a fleet like velocity of trawl over ground when trawling, 'hr' is the length of the head rope, 't' is the time spent, wing spread (hr × X2)
- ✰ Process: Estimation of swept area
- ✰ Output: Prediction of total biomass of a resource in a given area/coast.

Chapter 4

Biological Structure of Fisheries Resource in Space and Time

Fish ecologists and managers realize that all ecological systems exhibit heterogeneity and patchiness on a broad range of spatial and temporal scales. We all recognize that these patterns have fundamental effects on biological processes, dynamics of populations and ecosystems, our perception of the environment and, ultimately, how we sample and manage natural resources. Incorporation of space, time, and scale (and their interactions) in theories, modelling, and sampling has improved our understanding of how population dynamics and species interactions respond to the physical (*e.g.*, temperature) and biological (*e.g.*, prey density) spatial structure in the environment. Multi scale analysis has shown the existence of scale-dependent patterns that can be of biological or physical origin and that can have significant effects on biological processes. Recent advances in hardware and software technologies and analytical techniques have contributed to our abilities to acquire and analyse complex spatial data sets, and to model ecological processes, over a wide range of spatial and temporal scales. Despite these advances and our growing appreciation for the importance of space, time, and scale, it is still often ignored in field sampling and modelling programs that attempt to evaluate predator-prey interactions, fish growth and production, and sustainability of a fishery.

Strategies to be Followed

- ✰ Use contrasting modeling techniques to explore the potential mechanisms for small and large-scale horizontal migration/dispersal patterns and the corresponding patterns in spatial distribution at scales ranging from meters to thousands of kilometers and from hours to decades. The consequences of these migrations are evaluated in the context of foraging, predator-prey interactions, survival, recruitment, and exotic species invasions.

- Use Geographic Information System (GIS) technology to predict the landscape-level spatial distribution of fishes and explore the role of smaller-scale regional processes when predictions fail.
- Reduce variance in estimates of contaminant concentrations in fish tissue by considering spatial location of samples and identify distinct "hot spots" that may be responsible for contributing to high contaminant levels.
- Understand better the life stage dependent migratory behaviour of fishes in space and time.
- Use stable-isotope ratios of carbon and nitrogen from zooplankton and fish samples to identify seasonal and spatial trends in the source of energy for juvenile fishes, recruitment and nutritional processes in fishes.
- Explore the role of spatial process in fisheries-dependent data on our ability to interpret over-exploitation in a fishery. They provide an important lesson on how ignoring spatial issues in a fishery may have devastating consequences to fish populations.
- Explore how fine-scale processes affect the vertical spatial distribution of fishes and highlight the implications for foraging and growth. These contributions demonstrate the behavioural plasticity of fishes to changes in their local environment over dial and seasonal time steps, demonstrate and interactive effect of prey abundance and environmental conditions in mediating predator-prey interactions, provide insight into the spatial scale at which visual predators experience patchiness of prey, and provide relevance for the use of spatially explicit models of growth rate potential for quantifying pelagic habitat and changes in quantity and quality in habitat.
- Identify key processes controlling populations over various spatial and temporal scales.
- Use dimensionless ratios of rates to determine the relative importance of biological and physical processes across scales and demonstrate that dominant processes differ across various scales, life history stages, and species.
- Use a recently developed graphical technique to quantify how critical scales change with fish ontogeny.
- Develop computer-based tools for the analysis and exploratory comparisons of spatial data sets, patch identification, and quantifying spatial and scale-dependent proximity between predators and prey. These tools may be used for scale-robust hypothesis formulation and testing of spatial patterns of predators and prey.

Chapter 5

Indicators of Dynamics in Fishery Resource

Indicators are data or combination of data collected and processed for a clearly defined analytical or policy purpose. That purpose should be explicitly specified and taken into account when interpreting the value of an indicator. Fisheries indicators should provide practical and cost-effective means for the evaluation of the state and the development of fisheries systems and the effects that policy changes have on those systems.

In considering the concept of indicators of sustainable development, a necessary first step is to define what is meant by sustainable development in the context of fisheries. Sustainable development is generally defined as being development that meets the needs of the current generation without compromising the ability of future generations to meet their own needs. In the fisheries sector, the use of biological indicators in the development of fisheries assessments and management plans has been standard practice, in some of the countries for many years. However, relatively little attention has paid to the development of economic and social indicators that serve to assess progress on other aspects of sustainable development.

With respect to criteria for success it has been found, both theoretically and empirically that good indicators are easily measured, cost effective to collect, to calculate, and easily interpreted (to avoid confusion about the state of the system they are reflecting). Simple indicators are consistently found to outperform more complex (model-dependent) indicators, which are sensitive to data quality. There is consensus on the need for a suite rather than a single indicator and on the types of indicators that perform well regardless of system types.

Economic and Social Indicators

For the indicators to be effective and workable in assessing the economic and social performance of fisheries, they should have a clear policy relevance and in particular:

1. Provide balanced coverage of some of the key issues of common concern, and reflect changes over time
2. Be easy to interpret (that is, movements in each indicator should have clear link to overall sustainability)
3. Allow comparisons across countries
4. Lend themselves to being adapted to different national contexts, analyzed at different levels of aggregation and linked to more detailed indicator sets.
5. Be analytically sound in technical and scientific terms, based on internationally accepted standards and broadly accepted by stakeholders.
6. Be based on data that are available, of known quality and regularly updated.

Uses of Indicators

The main purpose in developing a set of sustainability indicators is to assist in assessing the performance of fisheries policy and management and to stimulate action to better pursue to sustainability objectives. This can occur in a number of areas. For example, indicators can be used for ex-post evaluations of the impacts of management initiatives, assessment of progress towards medium and/or long-term objectives and assessment of the impacts of fisheries.

They can also enhance communication, transparency, effectiveness and accountability in fisheries management. In this regard, indicators can be developed and reported at various levels of aggregation- international, national, regional and local levels. Many of the environmental indicators for fisheries referred to above are focused on the fishery level. Other aggregates that are regularly reported, such as the contribution of fisheries to exports, are reported at a national level. Yet others relate to fisheries that are managed regionally as straddling and/or highly migratory stocks.

Biological Indicators

Relative Biomass

Specifically of gelatinous zooplankton, cephalopods, small pelagic, scanvengers, demersal fish, piscivores, top predators, and biogenic habitat (cover forming species). Ease of sampling differs strongly between these groups (for instance external bodies, such as NGOs, often monitor the charismatic top predators, tourism can focus attention on coral reef fishes, snapshots of mangrove forests and shallow benthic habitat may be available from satellite images. It is possible to find proxies for even some of the most difficult groups. For example, changes in relative biomass of gelatinous zooplankton may be captured by frequency of bloom events, as this is a much more easily obtainable statistic due to the ability to generate it from expert information and fishermen interviews.

Biomass Ratios

In particular the biomass ratios of piscivore: planktivore (PS:ZP), pelagic: demersal (P:D) and infauna: epifauna. The last of these is probably not feasible in data poor situations, but the other two can be done.

Size Spectra

Which give an indication of perturbation in system structure (using the slope of the curve), but can also highlight changes in system productivity

Maximum (or mean) Length

This indicator is observed to work in practice even if simple rules of thumb regarding are used. Caution is needed regarding market driven changing in preferred sizes.

Total Fisheries Removals

This includes the main catch, incidental catch (by catch) and the part of the catch that return to the sea (discards). This indicator considers the total biomass removed from the system versus what is left cycling in the system. While similar in concept to the widely used comparison of primary production vs. removals from the system, there is concern that it will not be suitable for poor data and dispersed small scale fisheries due to the lack of data on removals let alone discards.

Diversity (Counts of species)

It remains an informative and fundamental piece of information about the system. Alternative measures of diversity or system structure may need to be considered too, depending on which operating model is used or what empirical data is available. One possibility is considering the value of changes in species-area curves (*e.g.* the slope and asymptote of the curve) through time (or spatially if under a perturbation gradient). It is likely that some simple or ordinal data can be collected on this from the fishers themselves given that they are acutely aware of what they catch. Looking at changes in these ranking could be highly informative.

Size at Maturity (Weight and length)

It is a strong means of detecting change in the system and stock structure. This size has to be monitored regularly. This is one of the strongest indicators of biological structure and suggests current level of fishing.

Biophysical

(Chlorophyll a, temperature, dissolved inorganic nitrogen, and level of contaminants): these may be drawn from water quality monitoring programs or remotely sensed data sets, but are needed even if they aren't already being collected as they are means of teasing out causation. Indicators respond to any system change, including those caused by sectors other than fisheries, so any information that helps elucidate causation is extremely helpful.

While all of these indicators are not equally easily calculated in data poor situations, they are a good "straw-man" to start testing indicators for such fisheries in this study. It is critical that a suite of indicators, which are not all highly correlated, is used;

Multiple time and space scales are spanned by the data sets.

Inclusion Criteria

It is also critical that data include species that:

- ☆ They are directly impacted
- ☆ They have high turnover rates, which may provide a noisy but early warning
- ☆ They define the habitat, as these often have a disproportionate or keystone role in the system
- ☆ They are from the upper trophic level, which are typically both vulnerable in their own right due to their life history characteristics, but also integrative of pressures and patterns at large scales.
- ☆ There is likely to be differential signal strength in pelagic and demersal system as pelagic systems are easier to characterize (require fewer parameters), but are more variable leading to longer periods before trends are detected. Concerns regarding the availability of extended time series at a single location may be addressed by considering snapshots at multiple different locations under differential perturbation (*e.g.*, fishing) pressure, so that it will still be possible to get a gradient and contextualization for any individual location. Caution may be necessary in such a situation though, as there may be problems if a changing environmental (or other) effect is masking the fishing effect.

Fisheries Indicators and Data Sources

The statistical information needed are:

- ☆ Total Allowable Catch (TAC), Quotas, by species,
- ☆ Quantity and value of landings, by species
- ☆ Employment (total; full and part-time; male/female; capture fisheries/aquaculture/processing industry)
- ☆ Fleet capacity (number of boats, by length or by GT/GRT)
- ☆ Quantity and value of aquaculture production by species,

- ☆ Recreational fisheries (estimates concerning the production and the number of people involved),
- ☆ Trade in fish and fishery products (by main species and main partners).
- ☆ Government financial transfers relating to the marine capture, aquaculture and processing sectors
- ☆ Status of fish stock

In broad terms, the Pressure State Response (PSR) framework aims to identify the pressure on the environment from human and economic activities, which lead to changes in the state or environmental conditions that prevail as a result of that pressure, and may provoke responses by society to change the pressures and the state of the environment.

Classification of some Indicators following the PSR Framework

According to the PSR framework that has been developed and used extensively, the above indicators can be classified as follow:

Dimensions	*Pressure*	*State*	*Response*
Ecosystem (resource and environment)	☆ Total catch ☆ Fish consumption	Stock status	TAC and quotas
Social	☆ Fishing effort ☆ Number of vessels ☆ Growth rate of number of fishers	Number of fishes	
Economic	☆ Subsidies ☆ Excess fishing capacity ☆ Profitability	☆ Profitability ☆ Sector employment	Economic incentives and disincentives (*e.g.*, subsidies, taxes, buy-back)

Chapter 6

Sampling Techniques

A section of population or otherwise called as sample is examined for the attributes concerted with the assumption that the selected sample is representative of the whole population. A sample is any number of units or objects selected to represent a population according to some rule or plan. Sampling is a method of selecting a fraction of the population in such a way that it represents the whole population. Good sampling requires large and long term investments interms of manpower and general expenses.

Sampling Fish

Length distributions of landed fish can be obtained from samples taken in the auction halls, ports, quays side sampling or at sea on commercial vessels. Generally, the length sampling is carried out by area, gear and time to ensure that the estimated size distribution is accurate. Age and weight-at-length can be obtained from these samples. Sampling throughout the year allows scientists to monitor differences in growth and age over the year. An unbiased sample should have fishes of a wide range of lengths and weights.

The sampled length frequency information for either landings or catch can be converted to ages using age-length keys derived from the relationship between age and length. Ages are estimated using otoliths. Counting the rings on these otoliths gives an estimate of the age of the fish. The numbers of fish at different ages for various gears, areas, quarters and nations are combined to produce annual catch numbers-at-age for each assessed stock.

Types of Sampling

Sampling Onboard the Fishing Vessels

Sampling for length composition of commercial catch should best be done on board of the fishing vessel during fishing operation. Because all fish caught are not always landed, smaller specimens are discarded at sea; it is of no market value. There are several practical difficulties with sampling programmes for collecting length data on onboard fishing vessels. It is difficult and costly to send scientists onboard in all selected vessels, even if the number of selected vessels is small.

Sampling at Landing Sites

Most convenient place for sampling fish for most fisheries is the site where fish are landed after capture, since a number of boats land at a landing site within a short period of time. It is possible for a single scientist to examine and measure sample from selected vessels.

The compositions of catch may sometimes vary between centre if vessels from these centre fish in different areas containing possibly of different stocks. The most practicable solution in this case would be to continue sampling at the convenient centre, but sampling may be done separately or combined after giving due weightage, according to landings from each area of fishing. When the fish are lying in a large heap on the beach, a small number has to be selected as sample. The usual tendency in the absence of a proper sampling system is to take the desired number of fish from the top of a heap and since it is a larger fish.

Most important raw materials required for the stock assessment studies are catch data, effort data and length frequency distribution of the landed fishes. All these data should be properly collected and tabulated for analysis. In India marine fish landings are monitored regularly by both the State and Central Departments. Multistage stratified random sampling is followed to assess stock. Normally it is recommended to collect data as prescribed below.

Number of Boats	*Sample Size*	*Recommended*
Less than 10	10	All boats have to be sampled
10-20 boats	10	One each for alternate boats
30 boats	10	One each for three boats
40 boats	10	One each for four boats
50 Boats	10	One each for five boats
100 Boats	10	One each for ten boats

Chapter 7

Age Determination in Fish

Age determination is one of the important tools in stock assessment and help to understand the dynamics of the population. The ability to estimate fish age accurately is essential in the study of the dynamics of fish population. The annual variation in a fish catch rely upon its growth pattern. It is often desirable to segregate the fish catch on the basis of age groups to understand the vulnerability of any specific group to the fishing gear. Age composition of the fish catch is being used in different fisheries of the world to predict the future available stocks.

The ability to perform age determinations based on the examination of hard anatomical parts is of fundamental importance in fisheries research. Certain structures in finfish and invertebrates taken from temperate waters show alternating patterns caused by seasonal changes in growth rates; a phenomenon similar to trees for which an age may be determined by counting annual rings in a cross section of the trunk. Validation of a regular periodicity in these marks permits assigning a time scale and determination of age. The successful application of techniques to enhance the detection of age marks in biological specimens is of vital importance in estimating growth and mortality rates, population age structure, and other parameters needed for understanding the dynamics of fishery resources and their response to natural phenomena and exploitation.

A wide variety of age determination techniques have been developed for finfish and invertebrates which depend on the detection of contrasting bands in body parts such as scales, otoliths, fin rays, spines, and vertebrae of fish, as well as external and internal structures of mollusc valves.

The age old method for determining the age of fish is analysing the length and weight. This method has application in culture fisheries only. Another method is tagging, clipping, releasing and recapturing and which is often used in capture fisheries. This is costly and often time consuming. Moreover, in developing countries this has some practical implications because of the non-availability of grants and ignorance amongst fishermen. Petersen's method requires large samples over long period. Moreover, this method is not applicable to partial spawner. Hence the only easily available method is that of using hard parts like scales, opercula, vertebrae, frontal bones, cleithra and fin ray sections.

Age assessment of fish from their calcified structures can be a vital component of most of our present day fisheries management decisions. Casselman (1987) coined the term 'Osteochronometry' or the estimation of the passage of time on age deciphered from bone-like or calcified tissue. Amongst the hard parts of a fish, scales, if present, can be removed from the body of the fish without causing any visible injury.

Fishes are known for indeterminate growth and which makes them an excellent material for study of age and growth. However, as 'old age' sets in, the growth is slow and often annuli are recorded close to each other and the measurements become difficult and sometimes confusing. Hence great caution should be exercised in reading the scales in older fishes.

The overall growth of the fish is directly or indirectly influenced by factors like food, inter and intra-specific or intrageneric competition, sex, temperature, density and productivity of water, the latter depends upon the prevailing physico-chemical conditions.

Reasonably standardized methodology for determining age in fishes were described by Mohr (1927, 1930, 1934), and Graham (1929) though work on this topic started in the eighteenth century (Bagenal and Tesch, 1978). Dahl (1909), Suvorov (1948, 1959), Hederstrom (1959) and Ricker (1975) described the history of age determination and growth in fishes. Buckmann (1929), Rounsefell and Everhart (1953), Lagler (1956), Parish (1956) and Chugunova (1963) described the methodology adopted in these studies. De Bont (1967) and Bagenal and Tesch (1978) sounded a note of caution, "Workers should proceed with caution, both with temperate and tropical fishes.

Carlander (1987) described the improvement in the methodology from the days of Van Oosten (1929) *i.e.* from cleaning the scales to their projection to get an enlarged image on ground glass. Microfilm readers are often used, though the results may not always be perfect. Recent use of sensitized screens connected to microcomputers increase the precision of measurement simplifying calculations (Frie, 1982).

(I) Indirect Methods of Age Determination

A. Age Determination using Hard Parts

(i) Scales

A random sample of fishes should be obtained from the commercial catch however, the composition of the catch depends on the efficiency of the gear. Scales collected from larger specimens may not clearly indicate the formation of either the larval mark or the first annuals may be indistinct. It is better to collect three to four scales from the region between the dorsal fin and the lateral line.

The scales for the study should be removed from the same place and such scales may be designated as 'Key Scales'. Mark four or five locations on the fish's body. Remove three to four scales from each location using large number of specimens from each size-group. Measure the lateral scale radius of each scale. Find the correlation coefficient between total fish length and lateral scale radius for each location. Only that location should be used for removing key scales which gives high value of correlation coefficient.

After each collection of scales, keep the envelopes in sunlight for about 5-6 hours in a basket to remove the moisture, otherwise fungus may develop on the scales.

Scales collected in the field are washed, cleaned and studied as dry mounts, after removing extraneous matter and mucous by washing them in tap water and rubbing in between finger tips. To make scales more clear and soft (in case of large scales), dip them in weak solution (1 per cent) of KOH for about 5-10 minutes, then wash in tap water and dry in air.

Small sized scales are mounted between two glass slides and studied with the help of compound microscope or stereoscopic binoculars at appropriate magnification, provided the eye piece is fitted with oculometer. In some cases the circuli are not clear in small scales. Under such situations, scales can be stained with Alizarine Red S and mounted in glycerine for study.

Large sized scales are read under the microfilm reader commonly known as "Dokumator". After washing, these scales are placed in between the glass strips and the magnified image is observed on the screen at appropriate magnification. Lateral scale radius and the distance between the focus and annuli are measured

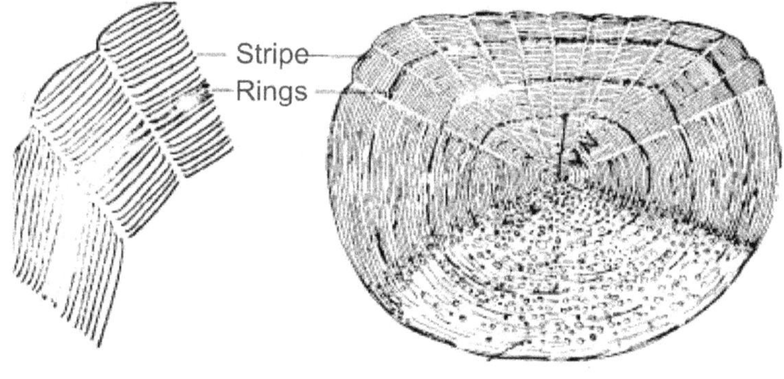

Figure 5: Scale of *Tilapia*

for studying relationship between scale radius/scale length and fish length. By extrapolating the regression line, correction factor is found out. The value so obtained is used for back calculations and other growth parameters.

(ii) Opercular Bones

It is easy to collect the opercular bones of the commercial fishes from the processing industries or fish market. The opercula should be detached from the skull and freed from extraneous muscles. These bones should be boiled in water for about ten minutes and all muscles may be cleaned with soft brush for 4-5 minutes. The annual rings become more clear with storage. Using black background, the opercular bones are studied under stereobinocular micro scope. Hansen (1978) recommended the study of opercular bones in glycerol after drying.

Due to large size of the operculum it is difficult to measure the opercular length and the annuli either under 'binocular or Dokumator'. Under such conditions trace the rings and mark them with pencil. Measure the opercular length O-A, the annuli O-A1.O-A2,O-A3,....O-An. Calculate the regression line between opercular length and fish length. Use the equation for back-calculations. There is no correction factor in this case.

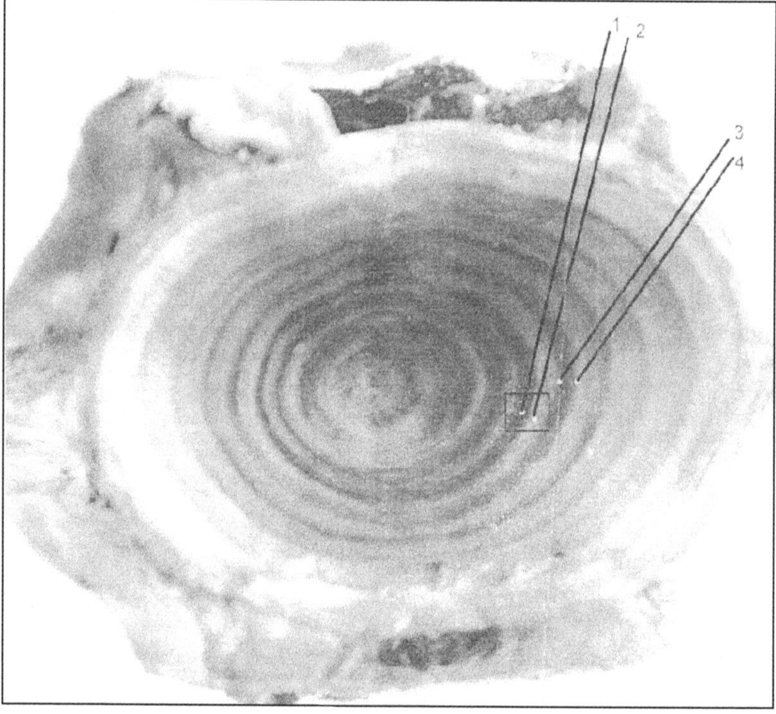

Figure 6: One Year's Growth in considered One Groove (3) and One Ridge (4). Stained rings are generally formed on the local side of the ridge (1) and Unstained on the Distal Side (2). (Report of the 2006 ICCAT workshop). Bluefin Tuna Vertebrae.

(iii) Vertebrae

The vertebrae lying below the dorsal fin (if two fins are present, then below the first dorsal) are removed and cleaned by boiling in water for 10-20 minutes to remove muscles. They are soaked for several minutes in detergent, rinsed in tap water and dried at 35°C. They can be stained with alizarine Red S (Galtsoff, 1952), rinsed and air dried. There are two methods for examining the annual layers.

a) The central point of vertebrae are examined under a stereoscopic binocular microscope.

b) The vertebrae are cleaved length-wise in the dorso-ventral direction. Half of the vertebra is fixed on wax so that the flat side is directed upwards and studied under stereobinocular microscope.

It is desirable to study large number of vertebrae for the clarity of rings before actually deciding the vertebrae to be used in future studies.

(iv) Frontal Bones

The bone is separated from the skull soaked in 10 per cent KOH for further cleaning. Bones are examined in 1:1 water glycerine solution under microscope. Growth rings are observed under reflected polarized light against dark back ground.

(v) Cleithra

Muscles are removed from cleithra either with a knife or by boiling them in water. The bones can be stored after drying in ordinary envelopes, or kept indefinitely in air with virtually no deterioration in clarity of optical zonation. These can be viewed under stereoscopic binocular microscope.

The use of cleithra and frontal bones in age determination studies is restricted.

(vi) Otoliths

Sagitta, Lapillus and Asteriscus are the otoliths present in the membranous labyrinth of fish on each side. Of these Sagitta is the largest and is often used in age studies, although the use of lapillus is also known. The otoliths exhibit best pattern of growth as calcium resorption is not known in them.

Otoliths removed from the membranous labyrinth can be stored in glycerol and water in the ratio of 1:1 with a little thymol added to the solution to prevent the growth of bacteria and fungus. Ethanol can also be used in place of glycerol. It has been reported that otoliths stored upto 5 years show no loss of clarity.

Dried otoliths are commonly employed for such studies. Their use has been found satisfactory in young and fast growing fishes. An otolith surface can be aged by immersing it in water and examining it under a dissection microscope using reflected light. Some otoliths have a cloudy or chalky surface rendering the identification of the growth zones difficult. The zones may be made slightly more distinct by rapidly dipping the otolith in a weak solution of HCL (usually 2 per cent) before placing it in water.

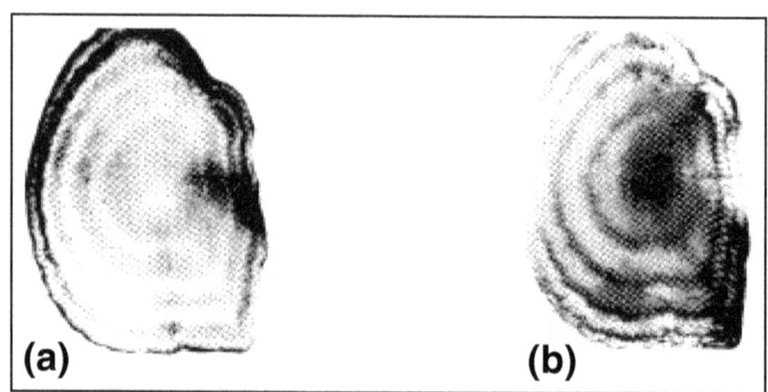

Figure 7: (a) Illuminated from Above Viewed against a Dark Background; (b) Illuminated from Below by Transmitted Light

The otoliths of slow growing fishes become thick and the yearly growth zones are not formed equally on all sides. To differentiate year marks, the study of cross sections is recommended.

Sagittae can be studied directly by placing them in alcohol, water or in dry condition. If the rings are not clear, grinding or sectioning is recommended. Otoliths are ground on the medial side down to the level of margin on 400-500 grade carborundum or stone or ground glass strip with 400 - 500 grade carborundum powder using dilute (1 per cent) HCl as wetting agent. Creosote oil is also used as wetting agent. The ground otoliths should be viewed under stereobinocular microscope. For sectioning, the otoliths are dipped in toulene to ensure that the epoxy will adhere to the structure. Sections are cut at 0.5mm thickness, or at any other thickness which yields best results. Several sections from one structure are placed on a slide with fast drying liquid or mounting material.

B. Age Determination Using Length Frequency Method

Determination of age and growth of a tropical species is more cumbersome compared to temperate species which exhibits seasonal spawning. Since the tropical system is more dynamic, further fluctuation in the environmental factors is common; the age of species gets affected. The food availability when abundance, when the physiological factors of fish are also at optimum, the growth of a fish is fast and as a result the attainment of maturity is early and more cohorts are released, in contrast to less food supply.

When more recruits are released particularly for a prolific breeder, assessing of age of fish by Length Frequency Data (LFD) will be erroneous by usage of conventional Petersen method.

Collection and Processing of Length Frequency Data

(i) Samples for length frequency should be collected at random from the commercial catches as soon as they are landed.

(ii) Sampling should be done before sorting into various market size groups.

(ii) Length data should be collected separately for different gears and for different mesh types of the same gear.

(iii) Length frequency data are to be recorded twice or thrice in a week from the landing centre.

(iv) While collecting the length data, sexes should be treated separately.

(v) Which recording the length data, total length or fork length or standard length should be taken depending on convenience.

(vi) The data should be recorded in mm or cm in a primary register.

(vii) It is better to take the length frequency data for a period of two years (24 months).

(viii) Classify the individual length recording into length groups which should not normally exceed twenty five.

(ix) The weekly data may be pooled on a monthly basis

(x) The data thus pooled must be drawn in the favour of frequency polygons or histograms for each month.

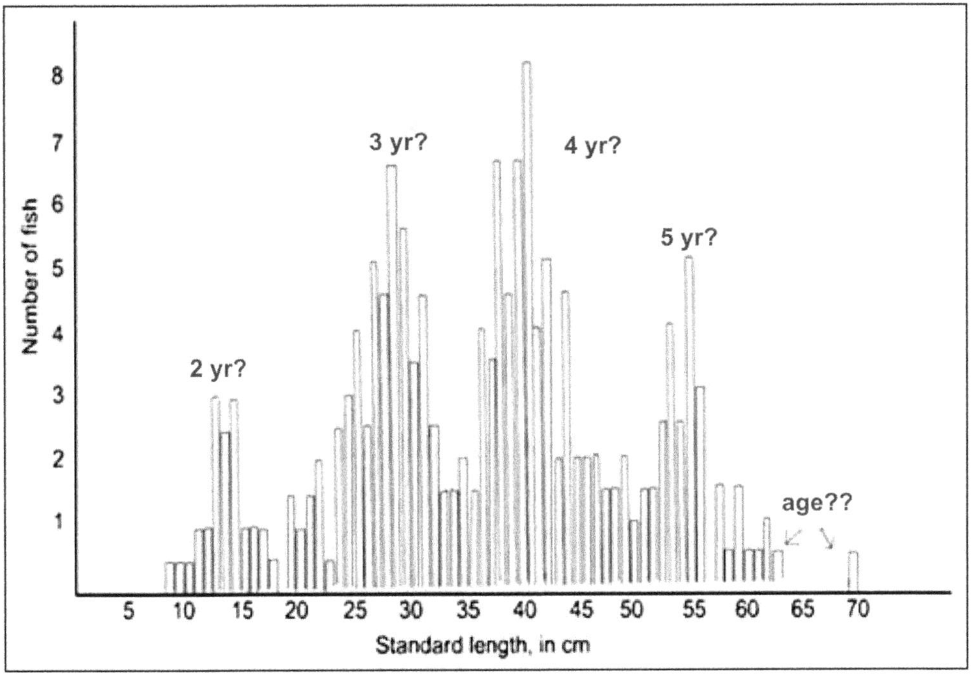

Figure 8

To ascertain the age of fish using LFD data, the most commonly used methods are

1. Petersen method
2. Modal class progression analysis
3. The integrated method of Pauly (Combination of first two methods)

1. Petersen Method

In this method, time intervals are taken into considerations in separating the various peaks of one length frequency sample. By this, the peaks are assumed to represent the different age group of a stock.

2. Modal Class Progression Analysis

For this method, LFD data should be taken for a minimum of two years, and the peaks are interconnected that belong to various samples arranged sequentially in time.

In length frequency method or size frequency method, the basic principle is (i) the length frequency distribution tend to group themselves around a central value called the mode and the progression of modes through successive intervals of time (*e.g.* month) indicates the pattern of growth.

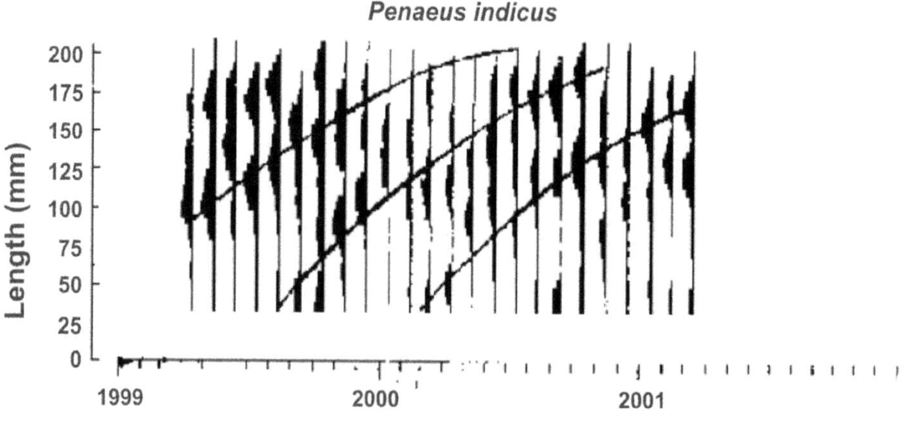

Figure 9: Month Mode Curve

3. Integrated Method (Pauly, 1983)

This method was devised by Pauly (1983) in which the above two methods are combined. The demerits of the first two methods are set right in Pauly's integrated method in which the growth curves are drawn by curved ruler directly upon the length frequency samples sequentially arranged in time. In this method, an attempt has been made to draw a growth curves with curved ruler directly upon the length frequency samples sequentially arranged in time. This methods is based on the following tenets (Pauly, 1983).

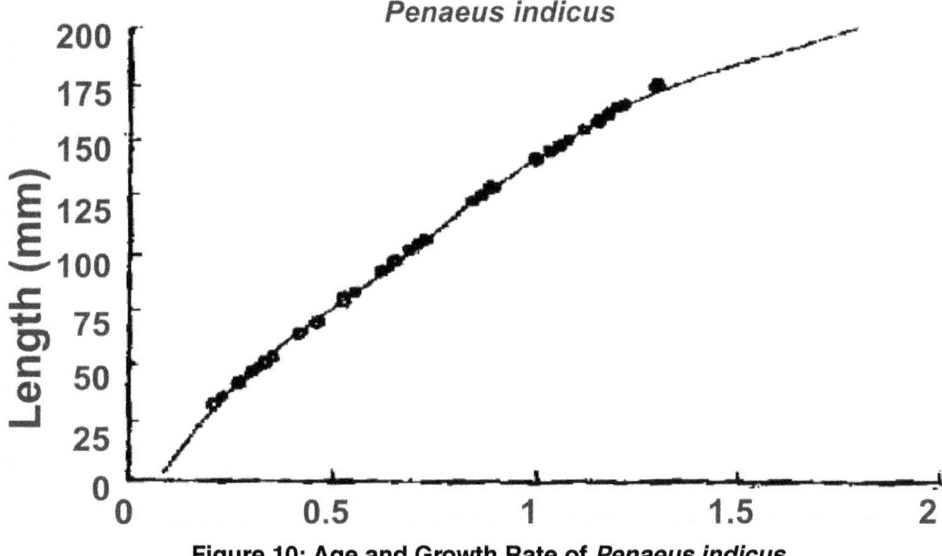

Figure 10: Age and Growth Rate of *Penaeus indicus*

Alternatively, plot the modes in the form of a scatter diagram against time in successive months and then trace their progression. See whether growth could be traced from the very origin of each brood to its maximum model position. and based on which plot age length key:

1. Length growth in fishes is at first rapid, then decreases smoothly and is, for the population as a whole, best approximated by a long continuous curve, rather than by several, short straight segments.
2. A single, smooth growth curve interconnecting the majority of the peaks of sequentially arranged length —frequency samples is likely to represent the average growth of the fishes of a given stock.
3. The growth patterns repeat themselves from year to year (which is also assumed when the "annuli" of otoliths are counted). While drawing curves, the following points are to be noted.
    i) The intervals on the time axis between the various samples must be proportional to the time elapsed between the sampling dates.
    ii) The original data must be plotted at least twice, or more along the time axis, which allows for longer, stabilized growth curves to be drawn and for all relevant age groups to be included in one single line.
    iii) When several growth curves are drawn (reflecting the production of several broods per year), the various growth curves should have the same shape, and vary only as to their origin.
    iv) The scale of the ordinate (length) should start at zero, thus allowing approximate spawning periods to be identified.

v) Each growth curve must interconnect several peaks. The more peaks a curve interconnects, the more likely it is to depict the actual growth of the population.

vi) The modal lengths corresponding to various ages (starting from an arbitrary age) can be read off the curve at regular time intervals, and may be then used to determine the growth parameters.

Chapter 8

Stock Structure Analysis

Population structure is of adaptive significance by enabling the species to exist under given conditions. To certain extent the population can adopt to changes in the environment. Even between species of aquatic fauna, differences exist. The age of population is controlled both by biological and ecological forces.

Laws of Population Structure

The size of a population at any given time may be taken to be a convenient index of many forces that act upon population. The growth of a population is determined by natality, mortality and dispersion. These are controlled by various ecological forces namely density dependent and density independent factors. Under optimal ecological conditions, a population in the absence of human intervention grows to its maximum size. This depends on the carrying capacity of ecosystem in which the fish lives.

Laws of Growth

Growth is an important quantitative aspect of development, the individual or a population grows and the growth show variations with the food supply. When a fish feeds actively, it undergoes maturation quickly and the rate of growth of an individual is also high. Rapid growth and larger size provide protection against enemies and escape from predators.

The increase in length before maturity is more dependent on the food intake than after maturity.

In principle, growth of fish is determined mainly by:

1. Conversion efficiency which mainly depends on physiology of metabolism and quality of food.

2. Trophic condition of the ecosystem in which the presence of favourite food sought by the fish.

Fish Growth in Tropical and Temperate Waters

Fishes in temperate waters live long compared to those in warmer regions. The growth coefficient, natural mortality are generally higher in tropical waters compared to temperate waters. Fish tend to mature according to a fixed portion to their asymptotic size (L_∞). Larger the asymptotic size, larger is also the size at first maturity. Hence for a given family of fish Lm/L_∞ is a constant ratio (Cushing, 1965). The asymptotic length and minimum maturity are generally higher in fishes of temperate waters. Because these two parameters are at higher side, it could be said Lm/L_∞ is inversely related to M/K. This is due to the inverse relation between K and L. In elasmobranch fishes of tropical waters, unlike teleosts grow slowly, live long and attain very large size. In general, elasmobranchs have long life with slow growth rate; the reproductive rate is also high with low reproductive stress.

Biological Features of Population

In a population, after it ages, reproductive potentiality decreases and avoidance of some spawning season is also common. The age structure of a population, its growth, sexual maturity changes with regard to the availability of food. This influences the size of fish and fecundity, as the number of eggs laid by a fish change with food supply and many recruits are released in a spawning season of a fish.

The growth in length and weight and fecundity of a population always follows sigmoid curve. In the initial life cycle, all the above biological factors will increase with age but once the population reaches a maximum size, the above factors shows decline.

Theory of Life Tables

The life table is a table of statistics relating to life expectancy and mortality for a given category of animals. Data on age specific mortality and fertility are organized in a life table. The life table has an important role in fish population. The structure of a population is specific of a species or subspecies and the environment. The population structure of a species or groups within species has certain stability. Once the environmental condition changes, the population of a species changes continuously, hence the structural resemblances and other features of the species show changes.

Life Span of a Fish

The life of fish varies greatly. About 75 percent of the fish species have life span between 2 and 20 years, 60 percent have life span of 5 to 20 years. Less than 10 percent have life span of over 30 years (Nikoloskii, 1980.) The largest fish with longest life span will usually have large predators. The life span of a fish stock is much influenced by food supply, maturation and even the number of recruits released. One form of the life table takes a cohort of individuals born at the same time and records the numbers remaining alive at regular time intervals until all dead.

Construction of Life Table

Construction of life table is a simple method, and it keeps the tracks of births, deaths and reproductive output of a population.

Three methods are used for construction of life table. They are:

1. Cohort Life Table

Cohort is a group of fishes that belong to the same age and share common growth and mortality parameters. They are normally recruited on the same day or spawned on the same spawning season. The cohort life table follows a group of same aged individuals from birth throughout their lives. This life table may work out for semelparous species, *i.e.* showing one reproductive event in their life time. This life table could be worked out easily by counting all individuals of a population during its life span.

2. Static Life Table

The static life table could be worked out for those organisms having complex life histories. This life table instead of working on a single cohort, it compares the population sizes of different cohorts of all life spans at a single point of time. The life table could be worked out with the assumptions that the proportion of individuals to each age class does not change and the size of the population nearly consistent.

Life table using mortality data need to be collected from a specified time period with a stable age distribution.

Both cohort and static life tables will be of less use to tropical fishes which shows year class variations and the pattern of recruitment is also not constant, as most of the species are prolific breeders. In such cases, the life tables could be worked out using mortality records.

3. Mathematical Models

Life tables are much needed in studying populations. To construct the models, the biology of the organism is an essential prerequisite. In addition to their biology, population may grow independent of their density, show exponential increase in number and are influenced by environmental factors. The population grow upto exponential rule but slow down when approach to carrying capacity (K). The 'r' is the intrinsic rate of increase per individual of a population. The 'K' is the maximum number of individuals of a particular habitat can sustain.

Equation for Density Independent Model for Organisms with Discrete Breeding Season

$$N_t + T = N_t + (T \times r \times N_t)$$

where,

N_t = Population size at a particular time.

T = time in years.

r = intrinsic reproductive rate

N_t+T = Population size to the subsequent year

Plotting Nt against time depicts the population changes each year under density independent conditions.

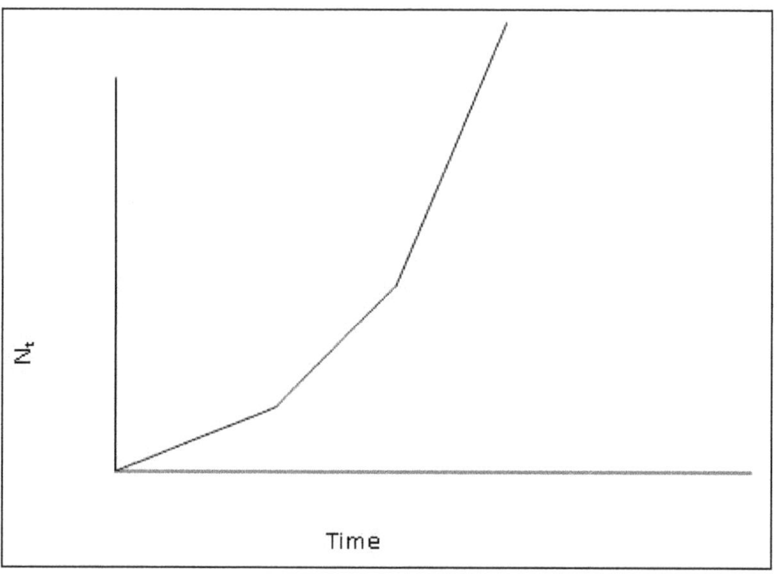

Figure 11: Exponential Increase of a population with Discrete Breeding Season (Density independent model for organisms with discrete breeding season)

Equation for Density Dependent Model for Population with Discrete Breeding Season

$$N_t + T = N_t + [T \times r \times N_t \times (K\text{-}N_t/K)]$$

where,

K = Carrying capacity

Once the N_t approaches K, the expression (K-N_t+/K) approaches 'zero' and N_t+T comes closer to N_t and at K, N_t +T equals N_t.

Plotting N_t against time shows increase for density dependent populations with discrete breeding seasons.

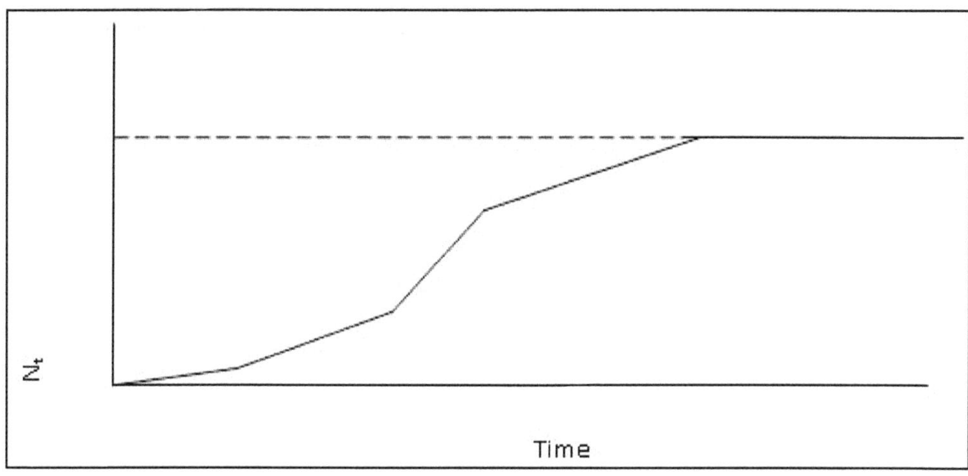

Figure 12: The Effect of 'r' and 'T' on Density Dependent Populations

Density Independent Model for Populations which Breed Continuously

The population tends to grow faster during the earlier period. If there is no effect due to density and environment, it will breed continuously. The equation for density independent model is

$$Nt = Nt\ e^{rt}$$

where,

Nt = The final population number

N_0 = The initial population size

e = The base of Natural logarithms.

r = Intrinsic reproductive rate

t = Duration

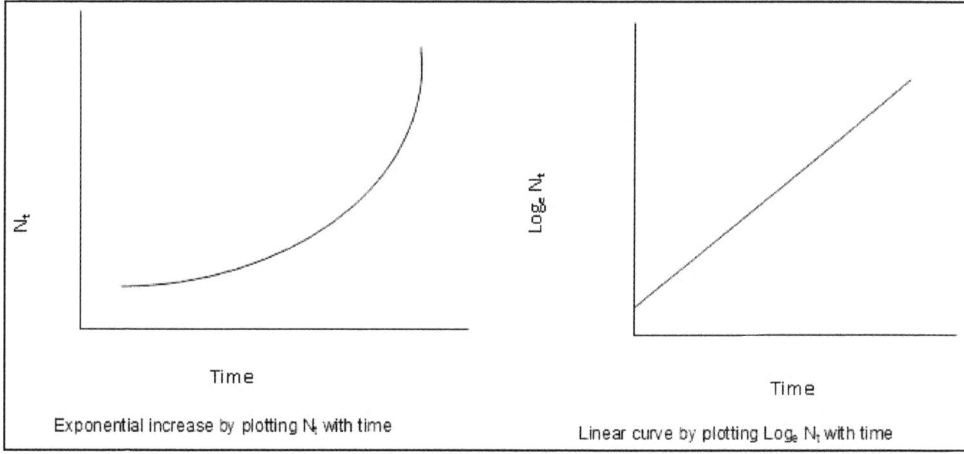

Figure 13

Density dependent model for population which breeds continuously

The equation is (density dependent model)

$$Nt = K/1 + e^{-(r_{max} \cdot t)}$$

Logistic population increase of the continuous density dependent model (N_t with time and attains asymptotic to K).

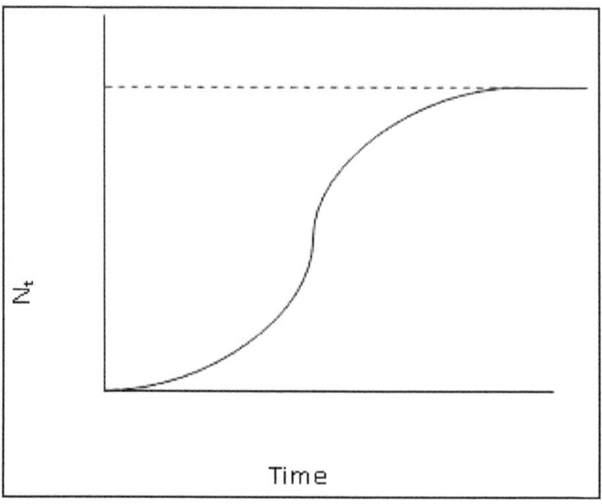

Figure 14

In the above graph, the population size follows the logistic pattern. It increases slowly and at a certain interval of time, it increases rapidly and once it attains K (the carrying capacity of the system) the line gets almost becomes stationary.

Virtual Population Analysis

The virtual population analysis (VPA) and cohort analysis are the methods used in fish population dynamics. This analysis is basically an analysis of catches of commercial fisheries obtained from fisheries statistics along with detailed information on the contribution of each cohort to the catch data.

In VPA analysis, once the history of population is known the future catches could be predicted.

The version of VPA suitable for pocket calculator is the cohort analysis. This analysis is identical to VPA and the calculation is also made simple. In VPA, the catch is taken continuously and in cohort analysis, the assumption is made that all fish are caught on one single day. The day is chosen to be I[st] July *i.e.* when one half of year has elapsed and natural mortality (M) is assumed to be constant throughout the life span of a cohort. The two methods namely Pope's cohort analysis and Jones length based cohort analysis are used for prediction of catches.

Cohort Dynamics

A cohort is a batch of fish of all of approximately the same age and belonging to the same stock. In the cohort of a stock, at birth the cohort has age zero. From age 0 to Tr, the cohort is in the "Pre-recruitment phase". The symbol N(t) is used to designate the number of survivors from a cohort attaining age 't'. The age is usually expressed in years. The symbol 'R' is used to designate the recruitment. 'N(Tr)' is the "number of recruits" to the fishery.

'Tr' is the minimum age at which the fish can enter the fishery ie., becomes liable to encounter with fishing gears.

'Tc' is called the age at first capture which marks the beginning of the exploited phase. 'Tc' is dependent on the mesh size.

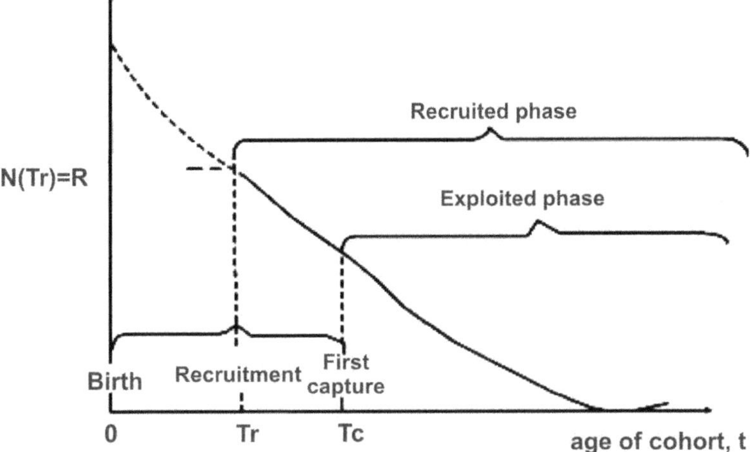

Figure 15: The Basic Features of Cohort Dynamics

Length based cohort analysis is available in computer programs namely "LCOHOR" in the LFSA package of microcomputer programs developed by Sparre, 1987. The FISAT (Gayanilo *et al.,* 1995) contain routines for a length based analysis similar to LCOHOR. In the above Figure 15, due to mortality there is a continuous decrease in the number of survivors. In the pre-recruitment phase (age '0' to Tr), there is only 'M' and after age 'Tr', the fish may be caught if suitable small meshed gear is used. At age 'Tc', the fish get caught with the mesh size actually in use.

Chapter 9

Growth Models

The study of fish growth and phenomena related to growth such as maturation, migration, food and feeding habits is a central theme in fisheries biology. The study of growth means basically the determination of the body size as a function of age. For fish stock assessment methods, the age composition data is of prime importance.

Growth

The simplest definition of growth of fish is that it is just an increase in size *i.e.* length or weight. During growth, the annual growth increment in length or weight steadily decreases through successive years of age while the length or weight itself keeps on increasing with age. The example given below demonstrates that the growth increment in length decreases continuously with the age of the fish.

Age in years (increase)	Length in mm (increase)	Interval between age	Annual increment in mm (decrease)
0	0	-	-
1	150	0-1	150-0 = 150
2	220	1-2	220-150 = 70
3	240	2-3	240-220 = 20
4	250	3-4	250-240 = 10

Growth Parameters

The growth parameters that are usually estimated in the fish stock assessment include:

- ☆ Length infinity or Asymptotic length ($L\infty$)
- ☆ Growth coefficient or curvature parameter (K)
- ☆ Initial condition parameter/Arbitrary origin of growth (t_0)
- ☆ Lmax
- ☆ Growth rate = $\Delta L/\Delta t$
- ☆ The growth increment during growth continues to decrease and attain virtually zero increment between two adjacent years of age in some future. The length of the fish corresponding to zero increment is called 'length infinity' or 'asymptotic length' and is designated by **L** ∞. This is the theoretical maximum length beyond which the fish cannot grow. In nature, the fish seldom attains this value but tend to grow towards it, as they die due to natural causes or due to fishing. **L**∞ is the mean length of the fish would reach if they were to grow to a very old age. It is one of the three parameters of the von Bertalanffy growth function.

Growth Coefficient or Curvature Parameter (K)

This is a parameter of von Bertalanffy growth function, expressing the rate at which the asymptotic length is approached. Short lived species, particularly, tropical fishes attain their $L\infty$ in a year or two and such species have high 'K' value. The long lived species usually have low 'K' value and need many years to attain their $L\infty$. Such long lived species have flat growth curve.

Initial Condition Parameter/Arbitrary Origin of Growth (t_0)

It is the initial condition parameter. "t_0" is the hypothetical age, the fish would have had at zero length if they had always grown according to the VBGF equation. t_0 should be read as t zero. Usually 't_0' has a negative value. t_0 is also called as 'arbitrary origin of growth'. Biologically 't_0' has no meaning because the growth begins at hatching when the larva already has certain length, which may be called L_0 when we put t = 0 at the day of birth.

L_{max}

L max is the length of the largest fish reported from a well sampled stock and T_{max} is the age of the largest fish reported from a well sampled stock. The L_{max}/T_{max} will be more accurate when the exploitation of the stock in question is low.

Growth Rate

- ☆ The growth rate is denoted by

(1)

- ☆ Time (or age), t is usually expressed in units of years.

☆ The mathematical relationship between the length of a fish and the growth rate at a given time is a linear function.

$$\frac{\Delta L}{\Delta t} = a + b^* L(t) \tag{2}$$

☆ Using VBGF equation, linear relationship can be derived as follows:

$$\frac{\Delta L}{\Delta t} = K^*(L_\infty - L(t))\ cm/year \tag{3}$$

where,

K = - b

L∞ = - a/b

☆ The mean growth rate (in length) can be calculated using the equation

$$\overline{L}_t = \frac{L(t+\Delta t)+L(t)}{2} \tag{4}$$

von Bertalanffy's Growth Equation

Length based von Bertalanffy's growth equation

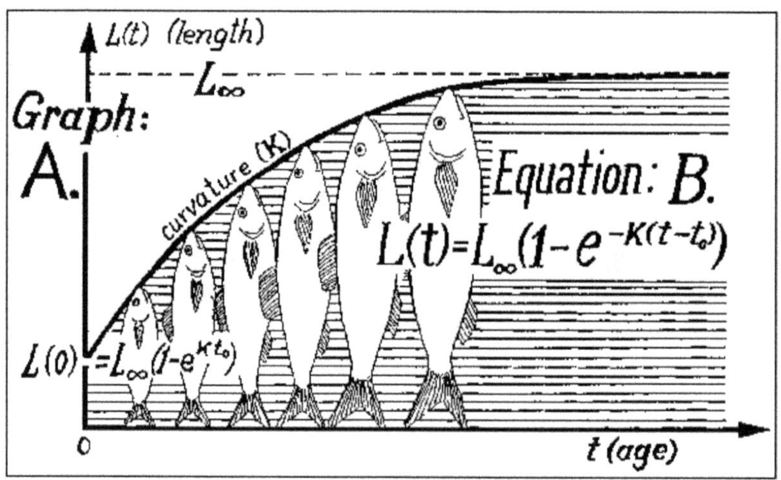

Figure 16

☆ Lt = L∞ [1 - exp (-K*(t – t_0)] (5)

where,

L∞ = Asymptotic length

K = Growth coefficient or curvature parameter

t_0 = Initial condition parameter or arbitrary origin of growth e = Base of natural logarithm

t = Time

Weight Based von Bertalanffy's Growth Equation

$$W(t) = W_\infty [1 - \exp(-K^*(t - t_0)]^3 \quad (6)$$

where,

W_∞ = Asymptotic weight

K, t and e and t are defined as above (Refer equation 5)

Determination of Growth parameters

Growth parameters are used as input data in the estimation of mortality parameters and in yield/recruit in fish stock assessment. Growth parameters differ from species to species, even between the sexes and also differ from stock to stock of a species. If differences exist, they should be calculated separately.

Growth Parameters and its Application

- ☆ Used as input data in estimating mortality parameters.
- ☆ Used as input data in the yield/recruit models in assessing the fish stock.
- ☆ Used to predict the relationship of temperate and tropical fish stocks.

Data Needed to Estimate Growth Parameters

- ☆ Length frequency data
- ☆ Mark recapture experiments (Tagging)
- ☆ Estimation of age and growth data by counting of year rings on hard parts such as scales, otolith sagitta or other bones

Methods/Equations Used to Study Growth Parameters

- ☆ Gulland and Holt Plot
- ☆ Ford-Walford plot and Chapman's method
- ☆ Chapman's method
- ☆ Bagenals' least square method

Gulland and Holt Plot

- ☆ The linear relationship could be derived from VBGF equation - cm/year

$$\frac{\Delta L}{\Delta t} = K^*[L_\infty - L(t)] \quad (1)$$

- ☆ This equation can be written as $\frac{\Delta L}{\Delta t} = K^* L_\infty - K^* \overline{L}(t)$ (2)

- ☆ (The length "L(t)" in the equation (2) represents the length range from Lt at age 't' to Lt + Δ T at age t + Δ T

- ☆ The mean length equation $\overline{L_t} = \frac{L(t+\Delta t)+L_t}{2}$ goes as an entry data in Gulland Holt plot.

- ☆ Using (L) (t) as the independent variable and $\frac{\Delta L}{\Delta t}$ dependent variable and above equation (equation 2) becomes linear function.

$$\frac{\Delta L}{\Delta t} = a + b^{*}\overline{L_t} \quad (3)$$

- ☆ The growth parameters K and L∞ are calculated by using the formula K = -b and L∞ = - a/b
- ☆ In Gulland and Holt Plot, the input data are't' and 'L(t)'.
- ☆ L(t) and Δ L(t) are calculated between the successive 't' and 'Δt' respectively.

$\overline{L_t} = \frac{L(t+\Delta t)+L_t}{2}$ is taken as 'x' variable, $\frac{\Delta L}{\Delta t}$ is taken as 'y' variable.

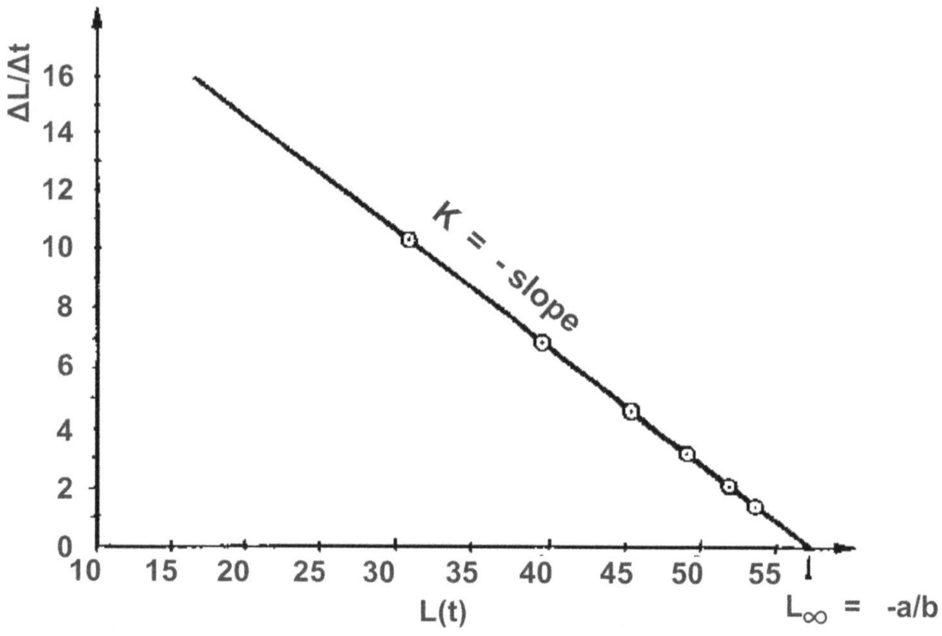

Figure 17: Plot of Growth Rate against Mean Length.

- ☆ Using the regression equation (y= a + bx), the K is determined by K=-b and L∞= - a/b
- ☆ A graph can be drawn by taking mean length as 'x' axis and $\frac{\Delta L}{\Delta t}$ in 'y' axis.
- ☆ The Gulland and Holt equation is reasonable only for a small values of Δt.

Ford – Walford Plot and Chapman's Method

- ☆ This method was introduced by Ford (1933) and Walford (1946). Without calculation, L∞ and K could be estimated quickly. The input data for Ford and Walford plot are L(t) as 'x' and L (t+ΔT) as 'y'. 'L∞' can be estimated graphically from the intersection point of the 45° diagonal where L(t) = L (t+ΔT)

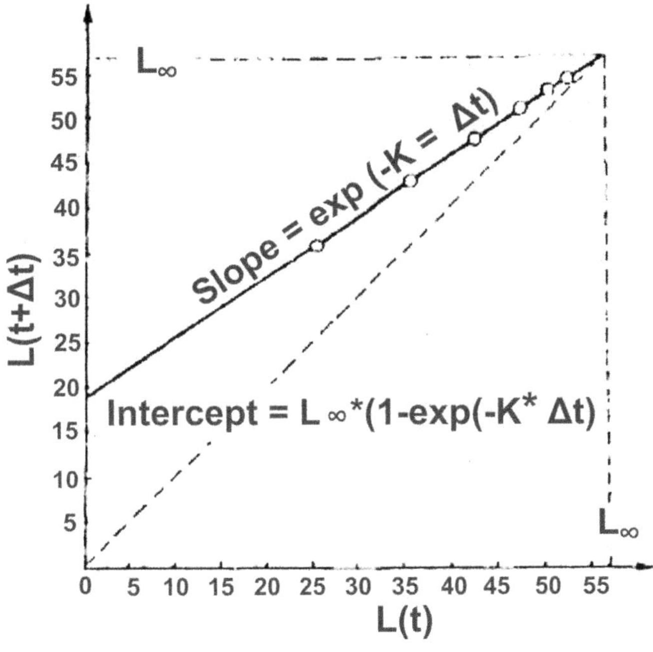

Figure 18

Chapman's Method

- ☆ The input data needed are L(t) and L (t+ΔT)- L(t)
- ☆ Note: The methods Chapman's and Gulland is based on a constant time interval ΔT if pairs of observation the methods could be used.

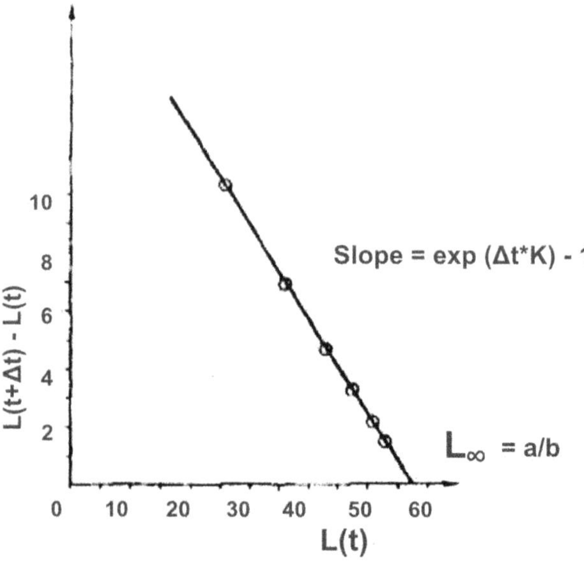

Figure 19

Bagenels' Least Square Method

- Growth in length
- The VBGF equation for growth in length is

  $Lt = L\infty\, [1 - e^{-K(t-t0)}]$ (1)

- The equation (1) can be rewritten

  $L_{t+1} = L\infty\, [1 - e^{-K(t-to)}]$ (2)

- The equation 2 gives a linear regression of Lt+1 on Lt of the type

  Lt+1 = a + b Lt (3)

  where $a = L\infty\, [1 - e^{-K}]$ and $b = e^{-K}$

  'a' and 'b' are estimated using the linear regression equation (y = a + bx)

Step (a): To estimate L∞ and Kn

- Transformation of length at age data into Lt and Lt + 1
- (Lt as x and Lt +1 as y)
- Apply linear regression analysis (y = a + bx)

Step (b): To Estimate t_0

- Take age in years as 'X' and $\log_e [L\infty - L_t]$ as 'Y'
- Apply linear regression analysis (y = a + bx)

- In the graphic method, 't_0' is written as:

$$t_0 = \frac{a - \log_e L_\infty}{-b}$$

- Which is of the simple linear form (y = a + bx)
- 't_0' could be estimated algebraically.

$$t_0 = \frac{1}{k}\left\{\left(\log e \frac{L_\infty - Lt}{L_\infty} + Kt\right\}\right.$$

The Weight-Based von Bertalanffy Growth Equation

- Combining the von Bertalanffy growth equation
  $L(t) = L\infty*[1 - \exp^{(-K*(t-t0))}]$
  with the length-weight relationship
- $W(t) = q*L^3(t)$
  gives the weight of a fish as a function of age

$$W(t) = q * L_\infty^3 * [1 - \exp(-K * (t - t_0))]^3$$

- The "*asymptotic weight*", $W\infty$, corresponding to the asymptotic length is

$$W_\infty = q * L_\infty^3$$

- The parameter, q, is called the "*condition factor*". Thus, "the weight-based von Bertalanffy equation" can be written:
  $W(t) = W\infty*[1 - \exp(-K*(t-t_0))]^3$

Growth in Weight

- The VBGF equation for growth in weight is
  $Wt = W\infty\ [(1- e^{-K(t-t0)}]^3$ (1)
- The equation (1) can be rewritten as

$$W_{t+1}{}^{1/3} = w_\infty^{1/3}(1 - e^{-k}) + e^{-k} w_t^{1/3} \quad (2)$$

- The equation 2 gives a linear regression of Wt + 1 on Wt of the type.
  $W_{t+1}{}^{1/3} = a + b\ W_t{}^{1/3}$ (3)

  Where, $a = W\infty^{1/3}\ (1- e^{-K}]$ and
  $b = e^{-K}$

Step (1): To find out 'a' and 'b' to arrive at $W\alpha$ and K.

- Applying simple linear regression to obtain 'a' and 'b' for the values of Wt (x) on Wt+1 in the data.

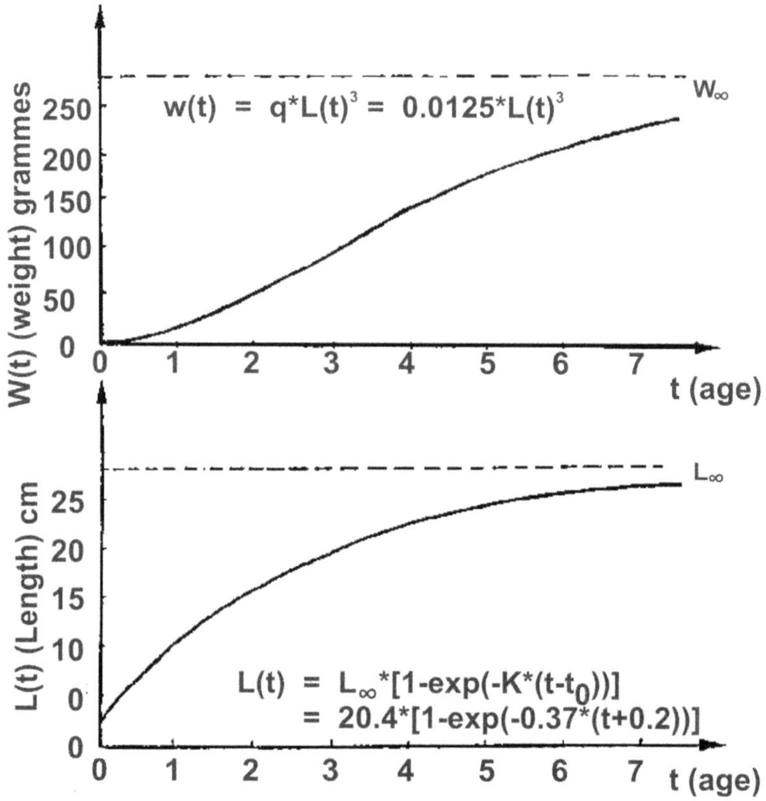

Figure 20: A Length-Based Growth Curve and Corresponding Weight-based Growth Curve

$W\infty^{1/3} = a/1-b$

$K = -\log e^{b}$

Step (2): To find out 't_0' graphically, take 't' as 'x' and $\log_e^{w_\infty - w_t}$ as 'y'

$$t_0 = \frac{a - \log_e w_\infty^{1/3}}{-b}$$

☆ To calculate growth parameters for weight based age data using von Bertalanffy equation, Wt and Wt +1 are converted to respective cube root equivalents. Once this is accomplished, the procedure mentioned in the different methods for the growth in length could be used for estimation of W∞, K and 't_0'.

Estimation of t_o

☆ The Gulland and Holt plot does not allow for estimation of the third parameter of the VBGF, 't_0'. This parameter is necessary. A rough estimate of 't_0' may be obtained from the empirical relationship.

$$\log_{10}(-t_0) = -0.3922 - 0.2752 \log_{10} L\infty - 1.038 \log_{10} K$$

Estimation of Growth Parameters for Elasmobranchs

☆ The growth parameters of Elasmobranchs

$$Lt = L\infty\, [1 - e^{-K(t - to)}] \qquad (1)$$

☆ This equation can be modified as

$$Lt + T - Lt = (L\infty - Lt)(1 - e^{-KT}) \qquad (2)$$

$$(i.e.\ Lt + T/L\infty = 1 - e^{-KT}) \qquad (3)$$

where,

Lt = Length at conception = 0 at zero time;

Lt +T = length at birth;

T = Length of gestation or hatchery period (the elasmobranchs being viviparous, ovoviviparous or oviparous, with the egg taking a long time to hatch).

L∞ = maximum observed length, *i.e.*, Lmax.

☆ Since there is evidence that 'T' is exactly the value of t_0 in Eq.(1) or its modification, Eq.(2) could also be expressed in the form,

$$Lo = L\infty\, [1 - e^{-K/(o - t0)}] \qquad (4)$$

where,

Lo = length at birth, (Lt + T) corresponding to 0 age, and t_0 = gestation or hatching period in years. Since Lo, t_0 and Lmax (= L∞) in the above equation can be empirically recorded, the only parameter to be computed is K.

☆ Once this is made, age for any given length can be estimated by incorporating K, t_0 and Lmax values in the equation (1)

Estimation of Growth Parameters for Crustaceans

1. Moulting is common in crustaceans. An individual crustacean usually will not obey von Bertalanffy's model but to some stepwise curve. Each step indicates a moult. However, the crustaceans moult different times in its life cycle. Therefore, the average growth curve of a crustacean will be a smooth curve.

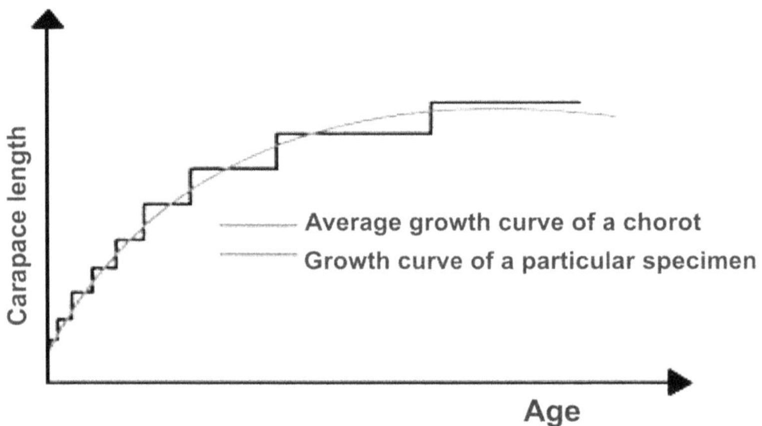

Figure 21: Individual Growth Curve and Average Cohort Growth Curve of Crustaceans

Growth Co-efficient

The growth parameter 'K' is related to the metabolic rate of the fish. Pelagic species are often more active than demersal species and have a higher 'K'. The tropical fishes have higher 'K' values compared to cold-water fishes.

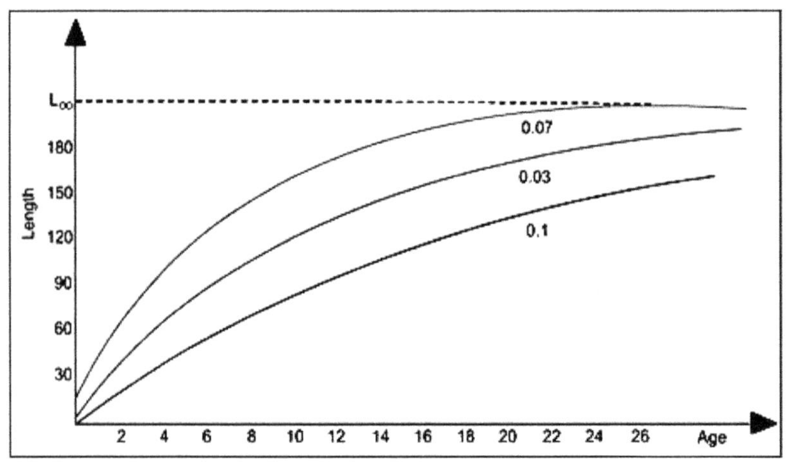

Figure 22: A Family of Growth Curves with different Curature Parameters, different k' values

The curvature parameter 'K' values are more related to 'M' values. In short lived species, particularly tropical fishes, 'K' is directly proportional to 'M'. The 'M/K' is inversely related to $L_m/L\infty$ and is also an index of reproductive stress. This value is found to be high for fish exhibiting post spawning death phenomena. (Lm is the minimum length at first maturity)

Temperate fishes live long compared to tropical species. The Natural mortality is high for short lived species.

$Lm/L\infty$ is an index of reproductive stress.

K is directly correlated with natural mortality.

$Lm/L\infty$ is inversely related to K and M.

(This is because, larger the asymptotic size, 'K' will be less and the rate of growth is high in short lived species, most of the tropical species have short life span compared to temperate species).

Chapter 10

Mortality Parameters

Fish mortality is a term widely used in fisheries science that denotes the loss of fish from a stock through death.

Types of Mortality

Natural Mortality

The removal of fish from the stock due to causes not associated with fishing. Such causes can include disease, competition, cannibalism, old age, predation, pollution or any other natural factor that causes the death of fish. In fisheries population dynamics, natural mortality is denoted by (M).

Fishing Mortality

It is the removal of fish from the stock due to fishing activities by any fishing gear. It is denoted by (F) in fisheries models.

Total Instantaneous Mortality

The removal of fish from the stock due to natural causes (M) and fishing activities (F). (M) and (F) are additive instantaneous rates that sum up to (Z), the instantaneous total mortality coefficient; that is, $Z=M+F$. These rates are usually calculated on an annual basis.

Application of Mortality Parameters

- ☆ Estimates of fishing mortality rates are often included in mathematical yield models to predict yield levels (Yield per recruit) obtained under various exploitation scenarios. These are used as resource management indices or in bio-economic studies of fisheries.

☆ They are used as input data in the Analytical Model of Beverton and Holt in estimating Maximum Sustainable Yield per Recruit. Further, this model predicts the effect of changes in fishing effort on future yields. In addition to the application in stock production modeling, they are also used as input data in cohort analysis, and in estimating standing stock biomass of aquatic animals.

Estimation of Total Instantaneous Mortality (Z)

Mortality parameters are extremely difficult to estimate precisely in aquatic animals. This is because of mortality rate, the rate at which the animals are removed from the population by death, is a parameter of population and not of individuals. To estimate mortality rate, it is therefore necessary to examine a group of aquatic animals.

Fishery scientists divide the instantaneous total mortality rate (Z) of a population into two parameters *i.e.,* instantaneous fishing mortality (F) and instantaneous natural mortality (M). An advantage of total instantaneous mortality is that they can be added or subtracted.

Thus, we have,

$$Z = F + M$$

where,

Z is total instantaneous mortality coefficient

F is instantaneous fishing mortality coefficient and

M is instantaneous natural mortality coefficient

In unexploited stock, F=0 then, Z=M. This means total mortality will have the same value when there is no fishing mortality.

Estimation of Z by Exponential Equation

In fishery biology, the most useful manner of expressing the decrease of the age group of fishes through time is by means of exponential equation. For example, if we put a group of fish in a pond and sampled the fish over a long period of time, the number of survivors of the original group decline over time as shown in the following figure.

This can be modeled in the following equation.

$$N_t = N_0 e^{-Zt} \qquad (1)$$

where,

N_t = Number of survivors after 't' units of time

N_0 = Original number of fish

e = Constant equal to 2.7182818

(e is the base of natural logarithm)

Z = Total instantaneous mortality rate in units of time^{-1} (*e.g.* 1/yr.)

N_t = The number of fishes remaining at the end of time 't'.

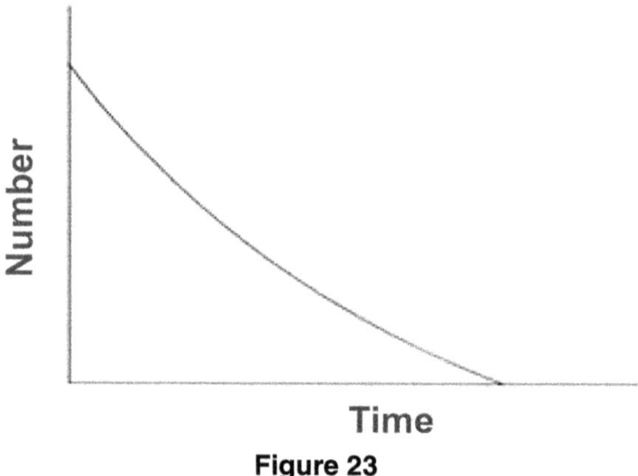

Figure 23

- Z' is most often reported in units of $(yr)^{-1}$. It is also reported in other units such as 1/month. To convert 'Z' to units of yr^{-1}, the equation (1) must be in units of months if 'Z' is in units of 1/month. If 't' is used as units of one year, then the number of time units is one twelfth (1/12) as large. (12 units of 1 month would become 1 unit of 1 year). So one should multiply Z by 12 in order for the answer (Nt) to stay the same. For example Z is 0.2/month, the Z in units of yr^{-1} will be 2.4 yr^{-1} (12 x 0.2 = 2.4 yr^{-1}).

Annual Survival Rate (S)

The proportion of the population that will survive after one unit of time has passed is known as annual survival rate (if time is expressed in years). The annual survival rate is denoted by 'S'. The relationship between annual survival rate (S) and the total instantaneous mortality rate (Z) is described by the following notion.

$$S = e^{-Z} \tag{2}$$

Annual Mortality Rate (A)

- The proportion of the population that does not survive is known as 'annual mortality rate' if time is expressed in years. It is designated by 'A'.

  $$A = 1 - S \tag{3}$$

- The annual mortality rate (A) and total instantaneous mortality rate (Z) will be more or the less the same to begin within a stock, but they progressively diverge as the mortality rate increases.

Estimation of 'Z' by Age Composition Data

Heinchke's Method

- This method is the oldest method. This method could be applied for short lived species (fish or shrimp or exploited aquatic animal) whose life span is more or less 2 years. Further '0' year class should constitute the fully recruited group.

  $A = No/\Sigma N$ (1)

- N_0 is the number of 0 year old fish and ΣN is the total number of fish comprising all age groups.
- From the annual mortality rate, S could be calculated as

  $S = 1 - A$ (2)

- Z could be estimated by the following equation,

  $S = e^{-Z}$. (3)

  $(- Z = \log_e S; Z = - \log_e S)$

Jackson Method

- In this method, 'Z', could be estimated between two adjacent age groups in a given year or between the abundance of the same year class in two adjacent years using the following formula.

  $S = N_{t+1/}N_t$ (4)

- Since the annual mortality rate (A) is the complement of S *i.e.*, A = 1 – S, 'Z' could be estimated by passing equations 3 and 4.

$$Z = \cdot\log_e \frac{Nt+1}{Nt} \text{ or } Z = \log_e \frac{Nt}{Nt+1} \quad (5)$$

Estimation of Z (if more than 3 age groups are represented in the fishery)

- If in a species more than 3 age groups (year classes) are represented in the fishery, the following formula is used.

$$S = \frac{N_1+N_2+N_3+N_4+N_5}{N_0+N_1+N_2+N_3+N_4+N_5} \quad (6)$$

$$Z = -\log_e S$$

Robson and Chapman's Method

- The 'Z' could be estimated from age composition data *i.e.*, the numbers caught per age group.

$$Z = -\ln \frac{A}{B+A-1} \qquad (7)$$

where,

A = N1 + 2*N2 + 3*N3 + 4*N4 + 5*N5+..................

B = N0 + N1 + N2 + N3 + N4 + N5+......................

Cushing Method

- ☆ When incomplete age composition data are available for a fish stock, for example only 2nd and 4th year age groups are available; Z could be calculated from the following relation.

$$Z = \frac{1}{3} \times \log_e \frac{N_2}{N_4} \qquad (8)$$

Estimation of Z from CPUE

Paloheimo (1961) Method

A plot of monthly CPUE against the respective months for a fishery prosecuted nearly all through the year may take the shape of a near normal curve, with the ascending phase including the peak falling in one half of the year and the descending phase in the other half. From this data, CPUE is derived separately for the descending phase (c/f)2 and ascending phase, (c/f)1, and Z estimated using the expression.

$Z = -\log_e (c/f)2/(c/f)1$

Estimation of Z using LFD and Selection Parameters

Four methods namely Ricker's Catch Curve, CPUE, Beverton and Holt equation and by Powell – Wetherall methods are used in the estimation of Z. In these methods, LFD and selection parameters are the data needed.

Estimation of Z by Means of a Catch Curve

This method is usually referred to as the length converted catch curve or linearized length converted catch curve. In this method, Z is estimated by sampling a multiaged population of fishes. Plotting the natural logarithm ($\log_e$) of the number of fishes in the sample (N) against their respective age (t) is called catch curve, and this is of the following equation.

$$\log_e N = a + bt \qquad (1)$$

Where, the value of b with sign changed, provides an estimate of Z.

Requirements to be Met by 'b' to be a Good Estimator of Z

According to Pauly, 1984, the requirements to be met by 'b' to be a good estimator of Z should include the followings:

- ✰ The values of $\log_e N$ should include the age group of fishes fully vulnerable to the gear in question and this corresponds to using only the descending part of the catch curve and it should be larger than the L'. (Smallest length of animals that are fully represented in the catch samples).
- ✰ All age groups used in the plot are recruited with the same abundance or the variations in recruitment fluctuations should be small and of random character.
- ✰ All age groups used in the plot should be equally vulnerable to the gear used for sampling.
- ✰ The samples used should be large and it should cover all age groups and it should represent average population structure over the period considered.
- ✰ Inclusion of fish whose size close to that of their asymptotic size must be avoided as this will result in their age being closely overestimated.
- ✰ Since 'Z' is equal to slope (with sign changed) of the catch curve, the real age which requires an estimate of T_0 can be replaced by relative age *i.e.* by setting 't' = T_0.

The equation - 1 can be rewritten as

$\log_e(N/\Delta t) = a + bt$

Where, Δt is the time needed to grow from lower (t1) to the upper (t2) limit of a given length class.

t = relative age corresponding to mid-range of the length class in question.

Input Data Needed

- ✰ LFD data and the catch in numbers in the respective length group.
- ✰ Values of L∞ and K needed for catch curve analysis, estimated from growth equations.

Estimation of Z by Growth and Selection Parameters (Beverton and Holt equation)

In this method, Z is estimated from the mean length of the fish in the catch and from von Bertalanffy's parameters ie., L∞ and K. Input data for the estimation in this method is less compared to length based linearized catch curve methods.

The equation $Z = K * \frac{L_\infty - \bar{L}}{\bar{L} - L'}$ (1)

$\bar{L}$ is the mean length of fish of length L' and longer. While L' is some length for which all fish of that length and longer are under full exploitation (According to Beverton and Holt, L' is the smallest length of animals that are fully represented in catch samples). K and L∞ are von Bertalanffy's parameters.

The equation is also be used based on length at first capture. This equation could also be written with the following notion. Here, $L_C < L'$

$$\text{Hence, } Z = K^* \frac{L_\infty - \bar{L}_c}{\bar{L}_c - L'} \qquad (2)$$

Lc is the length at which 50 per cent of the fish entering the gear are retained and $\overline{Lc}$ is the average length of entire catch. When $\bar{L}$ and L' are not available, the above equation (2) could be used for estimation of Z (when Lc and $\overline{Lc}$ are known).

Powell – Wetherall Method

In this method, plotting of $\overline{Lc}$ - L' against L' gives a linear equation from which 'a' and 'b' can be estimated, L∞ = -a/b; and slope b = -K/(Z+K) slope. This method is more suitable for situations where little or nothing is known about the fish stock in question.

Input Data Needed

1. LFD data
2. The number caught for each length interval are to be recorded.

Natural Mortality

Estimating natural mortality is one of the most difficult and critical elements of stock assessment. Natural mortality is denoted as 'M'. It will be high in larval and in early juvenile stages of fish. However, once the fish attain a certain age or size, natural mortality rate becomes much lower and much steadier and is generally considered to be constant. The von Bertalanffy's growth parameters, 'K' and 'L∞' are closely linked to M. As a rough generalization, fish species with a high K value have a high M value and fish species with low M will have low 'K' value. The ratio of 'M/K' value will be mostly in the range of 1.5 to 2.5 (Beverton and Holt, 1959). Natural mortality is linked to L∞ or the maximum weight of the species W∞, as the large fish have fewer predators than small fish. According to Rikhter and Efanov (1976), fish with high natural mortality mature early in life and compensates the high 'M' by starting to reproduce earlier. Natural mortality also has close relationship with the ratio of gonad weight to somatic weight in fishes. Thus, fish with high M may compensate by producing more eggs. As most biological processes go faster at higher temperature, M is also related to environmental temperature.

Application

This parameter is used in the estimation of yield/recruit (Beverton of Holt model) in fishes. This parameter also goes as an input datum in cohort analysis and in estimation of standing stock biomass of aquatic animals. If 'M' and 'Z' are estimated, 'F' could be estimated by the relation F = Z – M.

Estimation of 'M'

When values of 'Z' are available for several years pertaining to different annual values of effort (f), the value of 'M' can be calculated from

$Z = M + qf$

Where (q) is the 'catchability coefficient' or 'proportionality coefficient', which relates fishing effort (f) and fishing mortality (F). The fishing mortality (F) is almost always assumed to be proportional to effort and is mathematically expressed as,

$F = q^* f$

The more efficient the gear is, the higher the value of 'q' because 'q' is the measure of the ability of the gear to catch the fish.

When series of 'Z' (mean annual) values are plotted against their corresponding values of 'f', a straight line can be fitted to the points by means of linear regression technique. This results in a regression line with the equation

$y = a + bx$

The slope of the line is an estimate of 'q' and the 'y' the intercept is an estimate of 'M'. Sometimes, a regression line fitted to data of this type will give an estimate of 'M' which is negative. But 'M' cannot be negative. In such case, one makes an estimate of 'M' and then forces the regression line to have an intercept equal to 'M'.

If only one value of 'Z' and 'f' are available, or when there are only few values of 'Z' and 'f' are available, reasonable values of 'M' and 'q' could be obtained. In such case, the 'q' is estimated through

$$q = \frac{\bar{z} - m}{\bar{f}}$$

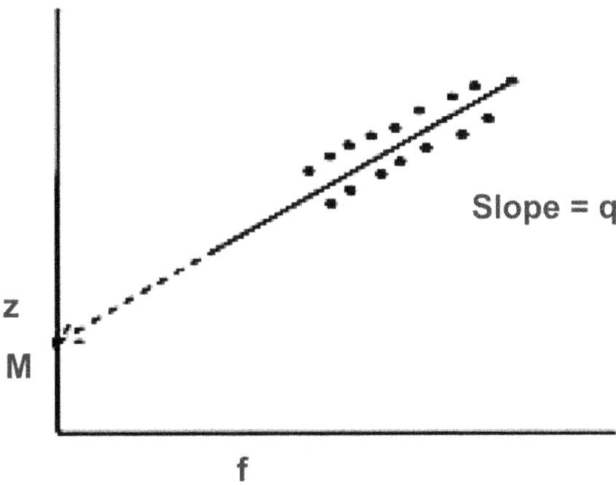

Figure 24

Where '$\overline{Z}$' is the mean of the available values of 'Z' (or a single value of Z) and '$\overline{f}$' is the mean of the values of 'f' (or a single value of f), 'M' being an independent estimate of Natural mortality.

Pauly's Empirical Formula

Natural mortality in fishes can also be correlated to mean environment temperature. The interrelationships can be expressed for length growth data by the multiple regressions as below:

For Length-Growth Data

$$\log_{10}{}^{M} = -0.0066 - 0.279 \log_{10} L\infty + 0.6543 \log_{10} K + 0.4634 \log_{10}{}^{T}$$

For Weight-Growth Data

$$\log_{10}{}^{M} = -0.2107 - 0.0824 \log_{10} w\infty + 0.6757 \log_{10} K + 0.4687 \log_{10}{}^{T}$$

'M' is the natural mortality in a given stock, '$L\infty$' (total length in cm) and '$W\infty$' (live weight in g) being the asymptotic size of the fishes of that stock and 'K' the growth co-efficient. The value of 'T', is the annual mean temperature in °C of the water in which the stock is found.

The above two equations give reasonable estimate of 'M' for about any set of growth parameters and temperature value. However for certain schooling pelagic fishes particularly for clupeoid fishes, the 'M' is generally overestimated by the above two equations. In such cases, it might be appropriate to reduce the estimate of 'M' by multiplying K by 0.8 (Pauly, 1983). Pauly's empirical equation should be avoided for crustaceans and molluscs or any other invertebrates, as the formula does not cover these groups.

Lorenzen (1996) proposed simple empirical method for estimation of natural mortality. The equation relates natural mortality with the body weight.

$$M = 3XW^{-0.288}$$

Petersen and Wroblewski (1984) provided an equation for natural mortality as a function of size - related to the theory on the distribution of biomass as a function of size:

$$M = 1.92\ W^{-0.25}$$

whereas Jensen related the natural mortality with the growth co-efficient and relationship is given as below.

$$M = 1.6K$$

Correlation of 'M' with Longevity of Fishes (Pauly's Method)

The growth co-efficient in fishes is closely linked with their longevity. In nature, the oldest fishes of a given stock grow to about 95 percent of their asymptotic size length. Thus,

$$Lt = L\infty(1-e^{-k(t-t_0)})$$

$$t\text{-}t_0 = \frac{\log_e(1-\frac{L_t}{L_\infty})}{-k}$$

If 95 per cent of L∞ is inserted for the oldest fish

$$t - t_o = \frac{2.9977}{-k}$$

Lmax ≈ 3/K

't max' is the longevity of the fishes in question.

Thus, natural mortality is correlated with size, since large fishes should have, as a rule, fewer predators than small fishes.

Rikhter and Efanov's Formula

Rikhter and Efanov (1975) showed a close association between 'M' and 'Tm 50 per cent', the age when 50 per cent of the population is mature (also called the age of massive maturation)

$M = 1.521/(Tm_{50\%}{}^{0.720}) - 0.155$ per year

'$Tm_{50\%}$' is equal to the optimum age defined as the age at which the biomass of a cohort is maximal.

Fishing mortality

Fishing mortality (F) is the rate at which the fishes are removed from the population by fishing activity. Fishing mortality is almost assumed to be proportional to fishing effort.

Estimation of 'F'

When 'Z' and 'M' are available, 'F' could be estimated from the relation

$Z - M = F$

In fully exploited stock, Z = F as 'M' is negligible and in unexploited stock 'Z' is an estimate of 'M'.

Exploitation Ratio and Exploitation Rate

Exploitation Ratio

The exploitation ratio is defined as the fraction of a year class recruits that is caught during all the year of existence (Ricker, 1975).

If M and F are available, exploitation ratio (E) could be calculated.

$$E = \frac{F}{Z}$$

This allows assessing whether the stock is overfished or not. According to Pauly (1983), on the assumption that optimal value of Eopt is almost equal to 0.5. According to Gulland the use of $E \approx 0.5$ as optimal value for exploitation ratio itself resting on the assumption that sustainable yield is optimized when $F \approx M$.

Exploitation Rate

The exploitation rate (U) is defined as the fraction of fish present at the start of the year.

$$U = E = \frac{F}{Z}(1 - e^{-Z})$$

where,

$$1 - e^{-Z} = 1 - S = A$$

Chapter 10

Gear Selection

The fundamental objective of any responsible fishing operation is to give maximum return to the fishermen with minimum efforts on the fish populations and the environment. Minimizing effects of fishing on the environment and fish population is much more difficult for quantification. This is because of complex interaction between the fishing process, fish and the environment. Over the years, improvements to fishing gear and fishing techniques have made fishing more effective with little consideration given to their effects on fish population and fish habitat. The effect of fishing on the population could be considerably reduced by making the gear selective.

Selection Process and Selectivity

According to Parrish (1963), selection in fishing can be defined as "any process that gives rise to differences in the probability of capture among the members of the exploitable body of fish". Such a general definition allows for the consideration of both between and within species selection during the different stages of the catch process:

1. In fact, the catch process (and hence selection) can be thought of as being divided in three distinct phases:
2. Probability that the occurrence of fish belonging to a single or different species coincides in time and space with the use of the fishing gear;
3. Probability that fish belonging to a single or different species encounters the fishing gear provided they are present when and where the gear is used (*i.e.*, that fish are accessible to the gear);

4. Probability that fishing gear retains fish belonging to a single or different species provided they have encountered it. (*i.e.*, that fish are vulnerable to the gear).
5. The first two phases are essentially dependent on fish distribution and behavioural patterns, while in the latter the specific characteristics of the fishing gear play a fundamental role.
6) When between-species selection is considered, capture will depend mainly on the behaviour displayed by each species towards the fishing gear, while in the case of within-species selection the retention of a fish will be driven by its specific characteristics (age, length or girth). In this case, selection is often taken as synonym of length selection, in spite the fact that where meshes are concerned, selection is essentially a girth/mesh-opening related process. Selectivity is no more than the quantitative expression of selection.
7. Unlike what happens for trawl codend selection studies gill-net selection studies have the drawback of the lack of knowledge on the structure of the population encountering the gear (with the obvious exception of the direct estimation studies). As a consequence, selectivity estimates are based on the comparative fishing with gillnets of different mesh sizes (the so-called indirect technique), while keeping constant the other physical characteristics of the gear. Furthermore, some basic assumptions are usually taken into consideration, the most important of which is the Baranov's "Principle of Geometric Similarity", which states that if selection depends only on the relative geometry of the fish and meshes then all selection curves are similar. Therefore, the selectivity will be the same for any combination of fish length and mesh size for which their ratio is constant (Hamley, 1975), that is to say that all meshes are equally efficient for the length class they catch the best.
8. Fishing selectivity can be defined as the ability to target and capture fish by species, size or sex during harvesting operations, allowing all incidental by–catch to be released unharmed.
9. Selectivity plays a major role in the development of a sustainable and economically viable fishery. The results of selectivity experiments can be befitted to the fishermen to allow them to capture only targeted fishes and ensures the essential return of juvenile fishes. Most fishing gears, for example, trawl gears are selective for the larger sizes, while some gears (gill nets) are selective for a certain langth range, only thus excluding the capture of very small and very large fish. This property of fishing gear is called "Gear Selectivity".

Bell Shaped Selection Curve

Bell shaped curve is important for the analysis of selectivity for most of fixed gears. The width of the selection curve provides the selection range of the gear

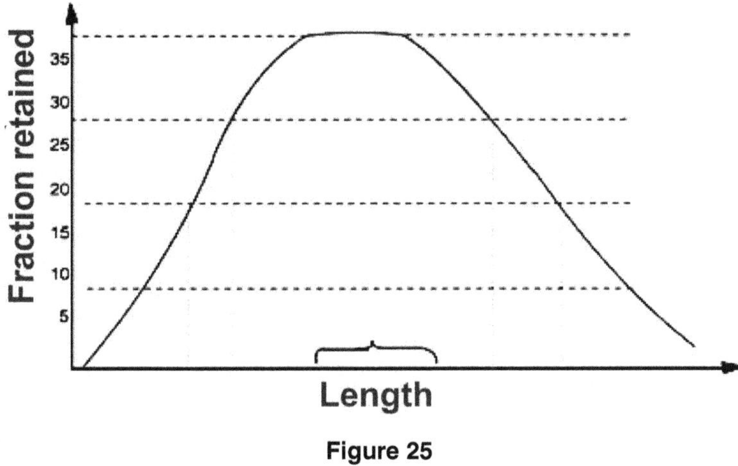

Figure 25

and the highest point in the curve corresponds to the optimum size of fish caught by the gear.

Sigmoid Shaped Selection Curve

In the fixed gear, the entry of larger fish is prevented but it permits the escapement of smaller fish in relation to size. By contrast, mobile gears allow all the smaller size fish to escape through meshes and larger fish become more susceptible to selection. As size of fish increase, the percentage of fish retained also increases. The selection process produces "S" shaped selectivity curve.

The point where 50 per cent of the fishes are retained and 50 per cent escape, is known as the 50 per cent selection size (L_{50}). The steepness of the curve indicates the efficiency of the selection access. A shallow curve that gradually changes between L_{25} and L_{75} depicts a poor selection access, whereas a curve that very quickly changes from 0 per cent to 100 per cent selection (a vertical line) indicates optimum selection. The small selection range produces a vertical line selection and this is said to approach "Knife edge" pattern.

Gillnet Selectivity

A gill net is made up of a long rectangular web of netting which is set in the water to a vertical wall. A buoyant head rope and a weighted foot rope ensure that the net wall remains in a vertical position.

Catching Process

Fish are caught in a gill net in one of three ways.

- ☆ **Wedging:** a fish is held tight around the body by a mesh.
- ☆ **Gilling:** a fish enters the net but it is too large to pass through and is prevented from retreating by twine that catches the fish behind its gills.

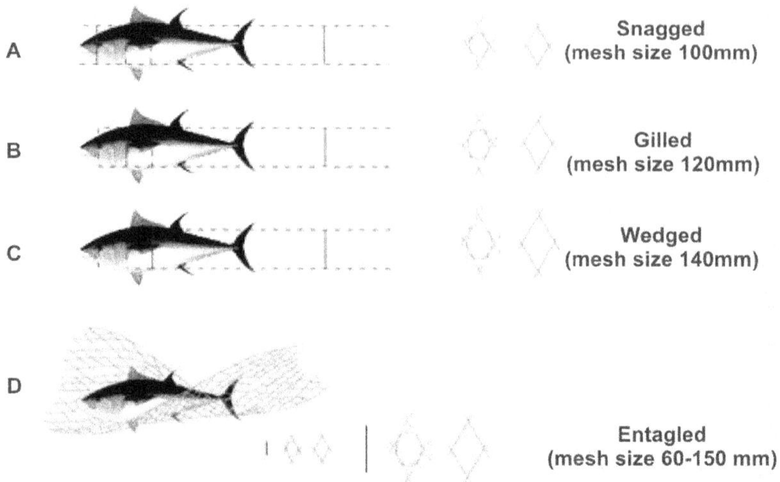

Figure 26: Gill Net Selectivity

- **Tangling:** The fish has not necessarily penetrated a mesh but is caught in the net by teeth, maxillaries or other projections.

Factors Influencing the Selectivity to Gill Nets

Planning an experiment involves prior knowledge of the factors that can affect gear selectivity. Some of them, as those related to the gear, are easy to control, in opposition to those related to the fish or to environmental conditions.

Gear Parameters

Gear and Net Dimensions

- Mesh size – The smallest fish caught has a maximum girth equal to the perimeter of the mesh, the largest fish caught has a head girth equal to the perimeter of the mesh. Fish between these two sizes are caught.
- Hanging ratio – the ratio between the head rope length to that of number of meshes multiplied by mesh size.
- Vertical slack (defined as the ratio between the stretched length of the inner and the outer netting, in trammels and semi - trammel nets). It can be controlled by the height of the walls in a trammel net.
- Twine characteristics (material, construction, thickness, colour, flexibility.)
- Floatation and weight
- Soaking time

- ✰ Arrangement of nets in the fleet - sequence and joining between nets; interaction between nets

Parameters Related to the Fish

- ✰ Fish abundance
- ✰ Fish availability to the net
- ✰ Fish behaviour towards the net
- ✰ Fish size
- ✰ Fish shape (girth at different body points)
- ✰ Presence of by-catch
- ✰ Presence of predators (can reduce the soaking time)
- ✰ Net saturation
- ✰ Patchy distribution in the net (includes attracting effects by individuals caught)

Parameters Related to the Fishing Operations

- ✰ Dimension of boats (low-lying vs. high-lying boats)
- ✰ Net handling techniques
- ✰ Environmental parameters
- ✰ Light level
- ✰ Sea state and currents
- ✰ Seabed type
- ✰ Depth
- ✰ Occurrence of water/bottom debris

General Information Needed

Before starting the work at sea, preliminary information must be collected regarding the fishery. Information on the following points should be collected:

Fish populations (species composition, distribution and abundance, seasonal variations, by-catch)

Fishing grounds (location, fishing yields, type of bottom, and presence of other static gear in the area)

Types of vessels used (main characteristics, bridge and deck equipment)

Net handling techniques

Duration of fishing trips

The main sources for this type of information are the fishermen themselves, the fish landings at local ports, or historic information from national databases.

Scientific information, if available, must also be collected regarding these parameters and others, such as:

Fish growth for the main species

Fish behaviour towards the gear

Characterization of the area in terms of environmental conditions

Estimation of Gill Net Selection Curves as per the Method of Sparre and Venema (1992)

Estimation of selectivity characteristics of gill nets requires the length frequency particulars from at least two or three gill nets of different mesh sizes with marginal difference. The mesh sizes selected should cover all the size group of aimed fish during fishing. The length frequency particulars can be collected from the gill nets operated by the fishermen or experimental gill nets fabricated as per our requirements. It is better to design the experimental gill nets by the researcher as it would help to avoid the experimental errors that may occur due to the variations in the designs of the experimental nets. Uniform hanging coefficient of 0.5 can be used in the experimental gill nets. The experimental nets should also have same dimension and similar sinking and buoyant force for comparison. Among different methods adopted for the estimation of selectivity characteristics, the method prescribed by Sparre and Venema (1992) is explained here. In this method, the natural log value of numbers of fish caught in gill net with mesh size 'A' and net with mesh size 'B' corresponding mid value of each length group are estimated as follows.

$$Y = \text{Ln}\ (CA/CB)$$

where,

CA = Fish caught in small meshed net A

CB = Fish caught in large meshed net B

Linear regression analysis is made against the mid value of each of the length group as follows

$$\text{Ln}\ (CA/CB) = a1 + b1L$$

where,

a1 = Intercept for AB combination

b1 = Slope value for AB combination

$$\text{Ln}\ (CC/CB) = a2 + b2L$$

where,

b1 = Intercept for CB combination

b2 = Slope value for CB combination

L = Mid length of a length group

CA, CB and CC refer to catch in number in net with the mesh size A, B and C corresponding to the mid value of length group, L.

The mesh size of gill net mainly depends on the girth of fish at the opercular region. For example, the mesh size of the experimental nets can be 9,11 and 14 cm for carangids and perches. These mesh sizes can be 26, 28, 32 mm for sardines/ lesser sardines.

The common selection factor (SF) for three-mesh combination *viz.* A, B and C can be derived using the equation.

$$SF = \frac{-2 \times [(A + B) \times a1/b1] + [(B + C) \times a2/b2]}{(A + B)2 + (B + C)2}$$

The common standard deviation(s) for three-mesh combination A, B and C can be derived using the following equation.

$$S = \frac{1}{31} \times \left[\frac{-2 \times a1 \times (B\ A)}{b21 \times (A + B)} + \left[\frac{-2 \times a2 \times (C\ B)}{b2\ 2 \times (B + C)}\right]\right.$$

The mean selection length (Lc) of carangids/perches in each net can be estimated from the common selection factor (SF) as follows:

$$Lci = SF \times mi$$

where,

'Lci' refers to mean selection length in different nets.

'mi' refers to different mesh size.

Lc for Net 'A' can be estimated by the relation

A Lc = SF x 9 cm

Lc for Net 'B' can be estimated by the relation

B Lc = SF x 11 cm

Lc for Net 'C' can estimated by the relation

C Lc = SF x 14 cm

The fraction of fish retained in the gill net (SL) can be estimated by the following relation

$$S_L = \exp\left[\frac{(L - Lc)^2}{2 \times S2}\right]$$

where,

Lc = Mean selection length

L = Mid length of the length group

S = Standard deviation for the three mesh combination

Estimation of Optimum Length of Capture

The commercially significant size group of the species under study in the fishery can be worked out as a percentage of catch from the pooled catch data of the gill nets of different mesh size used by the fishermen or from the trawl net catches of that fishing centre. Optimum length of capture can be fixed not only based on their significant contribution to the total catch but also based on its length at first maturity.

Estimation of optimum mesh size

The optimum mesh sizes for the commercial exploitation of the proposed species can be worked out based on the relation.

$$M = \frac{L_{opt}}{SF}$$

Where 'm' is the stretched measure of mesh in cm and Lopt is either the mid-length of the commercially significant length group or mean length at first maturity in cm. It is the responsibility of the fisheries manager to choose the correct Lopt based on the population dynamics and stock level of the species under study. The SF is the selection factor to be derived as per the method of Sparre and Venema (1992).

Trawl Net Selectivity

Trawl is a bag like net dragged on the sea bottom or through water column to capture fishes. Otter boards attached by long cables, keep the mouth of the net open horizontally during operation. Vertical opening of the net is achieved by floats and sinkers in the head and foot ropes respectively. The fishes are retained in the codend at the back of the net that consists of strong meshes.

Selectivity

Since the practice of trawling started, there have been concerns over trawling's lack of selectivity. Trawls may be non-selective, sweeping up both marketable and undesirable fish and fish of both legal and illegal size. Any part of the catch which cannot be used is considered by-catch, some of which is killed accidentally by the trawling process. By-catch commonly includes valued species such as dolphins, sea turtles, and sharks, and may also include sublegal or immature individuals of the targeted species.

Size selectivity is controlled by the mesh size of the "cod-end"—the part of the trawl where fish are retained. Fishermen complain that mesh sizes which allow undersized fish to escape also allows some legally–catchable fish to escape as well. There are a number of "fixes", such as tying a rope around the "cod-end" to prevent the mesh from opening fully, which has been developed to work around technical regulation of size selectivity. One problem is when the mesh gets pulled into narrow diamond shapes (rhombuses) instead of squares. The capture of undesirable species is a recognized problem with all fishing methods and unites environmentalists, who do not want to see fish killed needlessly, and fishermen, who do not want to waste their time sorting marketable fish from their catch. A number of methods to minimize this have been developed for use in trawling. Bycatch reduction grids or square mesh panels of net can be fitted to parts of the trawl, allowing certain species to escape while retaining others. Studies have suggested that shrimp trawling is responsible for the highest rate of by-catch. Selective trawl gear ensures the escape of small and/or unwanted fish during towing. Examination of this capture process reveals the importance of mesh shape, size and towing speed. Cod-end design is the selectivity of the gear.

Factors Affecting the Selectivity of Trawls

Mesh size and shape have been widely used as the traditional means of affecting selectivity in trawls. At high towing speed, diamond shaped mesh openings tend to become elongate and close altogether regardless of their size.

Methods of Measuring the Selectivity of Trawls

Covered Cod-end

To determine the size and distribution of the total fish population that comes in contact with the gear in a particular fishing area, attach a small mesh cover over the experimental net's cod-end. The small mesh cover retains the small fish that escape from the cod-end and so the combined catch of the gear and cover of the cod-end provides, a measure of the total population encountered. If series of different experimental cod-ends are performed with a small mesh cover, this gear design will enable the estimation of selectivity of cod-ends of different sized meshes and shapes.

Twin Trawl

Two trawl nets are dragged by a single trawler simultaneously. One trawl is composed of a small mesh cod-end used to obtain an estimate of the total population. The other trawl is used as an experimental gear. The selectivity of the test cod-end is determined by analysing the length frequency distribution of the catches from both the test cod-end and small mesh cod-end. Biasness is comparatively less in these methods.

Trouser Trawl

A standard trawl is divided into two sections by a vertical panel. Two cod-ends are attached, one on each side of the panel. Comparison of the length frequency distribution of the fish in each cod end allows calculations of cod end selectivity.

Alternate Hauls

This method requires identical hauls to be made alternatively with an experimental trawl and a control trawl. To ensure the validity and reliability of the data, each pair of hauls must be similar in all aspects and in the number of hauls made.

Parallel Hauls

This method requires that two vessels fish on the same ground at the same time. The experimental gear is towed by one vessel and the control gear is towed by the other vessel.

Estimation of Retention Fraction

Fraction of fish retained inside the net under each length group can be estimated as follows:

$$\text{Fraction retained } (S_L \text{ obs}) = \frac{\text{No. of fishes in the cod end}}{\text{Total number of fish caught both in cover and cod end}}$$

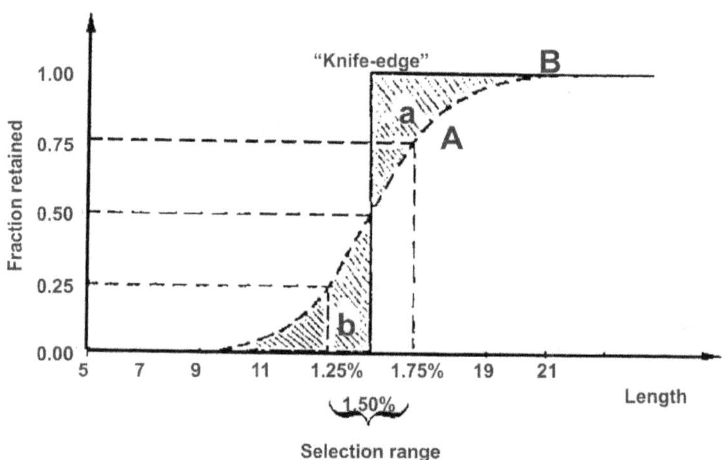

Figure 27: Trawl Selection Curve as a Function of Body Length. An illustration of the concepts of knife-edge selection and selection range

Trawl Selection Equation

The trawl selectivity can be defined by the following equation

$$S_L = \frac{1}{1 + \exp\,(S_1\, S_2 \text{ x } L)}$$

where,

$$S_L = \frac{\text{No. of fish of length 'L' in the cod end}}{\text{No. of fish of length 'L' in the cod end and in the cover}}$$

where,

L = Midpoint of length interval

S_1 and S_2 are constants

Linear Regression Analysis

The above equation can be rewritten as follows

$$Ln\ (1/S_L\ 1) = S_1\ S_2 \times L$$

This equation represent a straight line where

$S_1 = a$

$S_2 = b$

Ln $(1/S_L 1)$ is calculated for each length interval and filled in the column F which is treated as 'Y' axis.

Chapter 12

Surplus Production Models

Surplus production models are called as Holistic models. These models could be computed with less input data unlike analytic models. This model deals with total stock biomass along with fishing effort and yield. To operate this model, catch and effort data are needed as input data.

The objective of this model is to calculate the optimum level of effort. The total effort applied produces maximum yield. The maximum yield could be sustained by applying optimal effort without affecting the long term productivity of the stock. The yield thus generated without affecting the stock with optimal FMSY is usually referred to as Maximum Sustainable Yield (MSY).

Concept of MSY

Fundamental to the notion of sustainable harvest, the concept of MSY aims to maintain the population size at the point of maximum growth rate by harvesting the individuals that would normally be added to the population, allowing the population to continue to be productive indefinitely. Under the assumption of logistic growth, resource limitation does not constrain individuals' reproductive rates when populations are small, but because there are few individuals, the overall yield is small. At intermediate population densities, also represented by half the carrying capacity, individuals are able to breed to their maximum rate. At this point, called the maximum sustainable yield, there is a surplus of individuals that can be harvested because growth of the population is at its maximum point due to the large number of reproducing individuals. Above this point, density dependent factors increasingly limit breeding until the population reaches carrying capacity. At this point, there are no surplus individuals to be harvested and yield drops to zero. The maximum sustainable yield is usually higher than the optimum sustainable yield and maximum economic yield. MSY is based on a simple and easily understood concept of how a stock reacts to fishing. It is calculated on a simple physical measure. *i.e.*, weight of fish caught. In a steady state fishery, biomass declines as effort increases. The effort level corresponding to MSY being referred as FMSY. FMSY is used as biological reference point. MSY and FMSY could be estimated by surplus production models and by prediction models.

Surplus Production Model

This model does not take into account age and growth. Hence, it could be safely applied to tropical stocks, where calculation of age of tropical fish is more cumbersome.

When catch and effort data are applied for a number of years, the MSY thus generated will be more meaningful for sustainable fishery.

The two models will be highlighted in this chapter.

1. Schaefer model
2. Fox model

1. Schaefer Model

- ☆ The catch per effort or yield per unit of effort is designated as Y/f. The Y/f is a function of effort, 'f'. MSY could be computed from the following equation.

  Y/f= a + b x f(i) (1)

- ☆ In Schaefer model, the slope 'b' will be negative if the catch per unit of effort 'Y/f', decreases for increasing effort. This model implies one effort level for which 'Y/f' value obtained just after the first boat fishes on the stock for the first time. Hence, the intercept value is positive. Thus '- a/b' is positive and 'y/f' is zero for f = -a/b. As the negative value of catch per unit effort, 'y/f' is will not be a reality, this model applies to f values lower than '- a/b'. For unexploited stock or for stock which is less exploited, this model could be used. Even at very few levels of effort, the straight line reaches zero and attains annihilation.

2. Fox Model

- ☆ This model gives a curved line when 'Y/f' is plotted directly on 'f'. But a straight line is obtained when the logarithms of 'Y/f' are plotted on effort.

  Ln [y(i)/f(i)] = c+ d *f(i) (2)

- ☆ The above equation could also be written as

  Y(i)/f(i) = exp. (c+d *f (i))

☆ Schaefer model plot of 'Y/f' on 'f' gives a straight line and in Fox model plotting of 'Y/f' on 'f' gives a curved line which approaches zero only at very high levels of effort without ever reaching it.

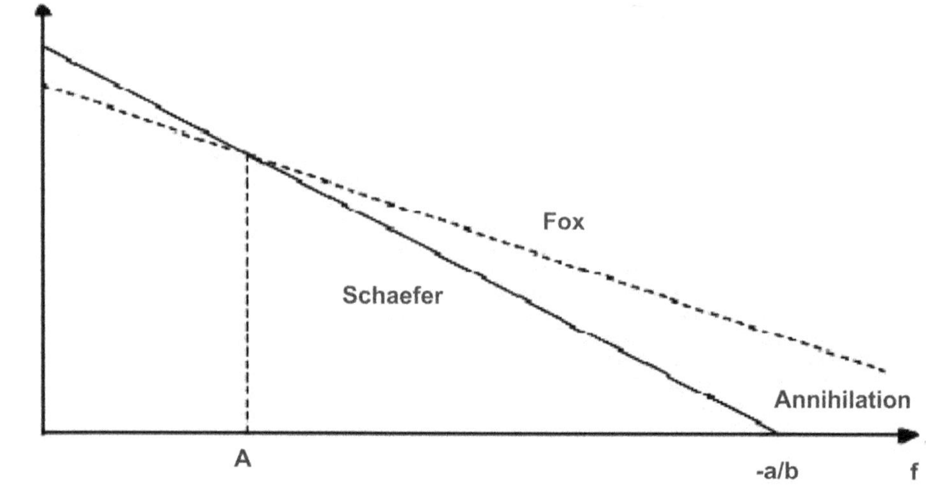

Figure 28

☆ The Figure 28 depicts another basic feature of the two models. When effort (f) is close to zero, 'Y/f' takes the maximum value and also the biomass, because Y/f = q. B and 'q' is constant. The biomass corresponding to f = 0 is called virgin biomass or the unexploited biomass. This is denoted by 'Bv'. In the equation of Schaefer and Fox, the 'Y/f' can be replaced by 'q'. 'Bv'. as follows:

q * Bv = a or Bv = a/q (Schaefer)

q * Bv = exp. (c) or Bv = exp.(c)/q. (Fox)

☆ In both models, the unexploited biomass (Bv) is the same. When the effort (f) is increased to level of A (see above Figure 28), the two curves (Schaefer and Fox curves) are approximately equal. But increasing the effort to the right of A, the differences become larger. Thus, one can use any one model and it becomes important only when relatively large values of 'f' are reached.

Procedure

(i) Collect data on yield of a stock.

(ii) Collect effort data (Number of standard vessels) employed to catch the stock in question.

Collect these two data for a number of years. For Schaefer model, effort f(i) is taken as X and Y(i)/f(i) (catch per unit effort) is taken as Y.

For Fox model, effort f(i) is taken as X and ln [Y(i)/f(i)] is taken as Y variable.

In Schaefer model,

$MSY = -0.25 . a^2/b$

$FMSY = -0.5 * a/b$

In Fox model,

$MSY = -(1/d) * Exp. (c-1).$

$FMSY = -1/d.$

(Note: The constants 'a' and 'b' for Schaefer model should be replaced by 'c' and 'd' in Fox model)

Uses of the Surplus Production Model

The advantage of surplus production model is that it can also be used to estimate a first approximation of the sustainable yield from fisheries operating on multiple stocks in a well-defined area, treating them all as a unit stock. Careful examination of time series data on catch and effort would help in identifying the successive phases in the growth of a fishery from its origin to peak and then down to decline. These phases could be indicated in the relative and absolute yield curves by appropriate identification marks for each phase so as to make the model more meaningful and readily usable as a tool for management and development decisions. This model will also serve as an excellent tool for ready estimation and judicious allocation of optimum yield and effort to solve intersectoral and interstate conflicts which are very common in the Indian fishery industry.

Relative to the Schaefer model of surplus production, the Fox model seems to be more realistic when a fishery undergoes transition from traditional to mechanized state or when mechanized fleet is added on to the existing traditional fleet, when there are changes in yield and effort in quantitative and qualitative terms.

Merits and Demerits of MSY

Merits

MSY has been especially influential in the management of renewable biological resources such as commercially important fish. Maximum sustainable yield (MSY) is the largest average catch that can be captured from a stock under existing environmental conditions. MSY aims at a balance between too much and too little harvest to keep the population at some intermediate abundance with a maximum replacement rate.

Demerits

The main demerits are usually at MSY, the catch rate, mean size of fish, and biomass are too low, and sometimes the recruitment may become less regular as FMSY is approached. Because of this, there will be poorer recruitment than average recruitment year. This will have a significant impact on fishing industry. In such cases, if total allowable catch is practiced by a quota, there will be chances of heavy overfishing.

According to economists, the physical maximum yield does not make economic sense because the total catch increases very little with a greater increase in fishing effort near the maximum. There will be high cost in effort with relatively little increase in yield.

FMSY and FMAX

FMSY is the effort level corresponding to MSY differs in that of FMAX. FMAX is fishing mortality rate at which average catch per recruit is at a maximum. FMAX is in weight. It is a function of growth rate and mortality rate of the stock as well as the size at which the fish enter the fishery. FMSY is independent of recruitment.

FMSY refers to fishing mortality rate at which the average long term catches from the stock is the highest. Hence, FMAX is a function of total production increase within the stock including efforts of changing yield for recruit and recruitment, which is also affected by fishing. In Maximum Economic Yield, the catch at this level will be less than open access catch on MSY by the economic rent to the industry *i.e.* surplus production will be at its maximum.

Limitations of MSY Approach

Although it is widely practiced by state and federal government agencies regulating wildlife, forests, and fishing, MSY has come under heavy criticism by ecologists and others from both theoretical and practical reasons. The concept of maximum sustainable yield is not always easy to apply in practice. Estimation problems arise due to poor assumptions in some models and lack of reliability of the data. Biologists, for example, do not always have enough data to make a clear determination of the population's size and growth rate. Calculating the point at which a population begins to slow from competition is also very difficult. The concept of MSY also tends to treat all individuals in the population as identical, thereby ignoring all aspects of population structure such as size or age classes and their differential rates of growth, survival, and reproduction. As a management goal, the static interpretation of MSY (*i.e.*, MSY as a fixed catch that can be taken year after year) is generally not appropriate because it ignores the fact that fish populations undergo natural fluctuations (*i.e.*, MSY treats the environment as unvarying) in abundance and will usually ultimately become severely depleted under a constant-catch strategy. Thus, most fisheries scientists now interpret MSY in a more dynamic sense as the maximum average yield (MAY) obtained by applying a specific harvesting strategy to a fluctuating resource.

Maximum Sustainable Yield and Maximum Economic Yield

In population ecology, Maximum Sustainable Yield (MSY) or surplus production is the highest theoretical equilibrium yield that can be continuously taken (on average) from a stock under existing environmental conditions without affecting significantly the reproduction process. The MSY is also called as Potential yield.

Maximum Economic yield (MEY), is the value of largest positive difference between total revenues from and total costs of fishing with all input valued at their opportunity costs when relating total resources from fishing to total fishing effort in surplus production model. It is also called as Economic Optimal Catch.

Swept Area Method

For any analytical model of fish stock assessment, qualitative and quantitative database is an essential prerequisite. Holistic models are recommended for the situation when there is limited scope for collecting huge data for particular fishery. Holistic model is simple, less data demanding, and also they do not require age or length structures of the stocks, but consider stock as a homogeneous biomass. The basic component of this models is the collection of data over a period of time along with effort or swept area. Two methods followed in holistic methods are 1) swept area method and 2) surplus production model. In swept area method, it is assumed that the area swept by the trawl is representative for the entire area over which the fishes are distributed. It is based on research trawl survey catches per unit area. From the densities of the fish caught in an area where fishing has been done. From the data the maximum sustainable yield is calculated.

Trawl net is a conical bag with a wide mouth fitted with otter boards for horizontal opening and floats and weights for vertical opening. This net sweeps the sea bed over a wide area. The tail end of the gear where the fishes are collected is called "cod-end". Trawl nets dragged on the sea floor are called "bottom trawls". The mean catch (either in weight or in numbers) put unit of effort or per unit of area is an index of the stock abundance. This index is converted into an absolute measure of biomass. This technique is called "swept area method".

The trawl sweeps a well defined path, the area of which is the length of the path, times the width of the trawl called "Swept area" or "the effective path swept".

The swept area 'a' can be estimated by following formula

$$a = v * t * h * x2$$

where,

V = Velocity of the trawl over the ground when trawling

h = Length of head rope

t = Time spent for trawling

x2 = Wing spread as fraction of the head rope length (0.4 - 0.6)

When exact position of the start and end of the haul are available, the distance covered can be estimated by

$$D = 60 \text{ x } \sqrt{(Lat1-Lat2)^2+(Long1-Long2)^2.\cos^2(0.5x(Lat1+Lat2))}$$

where,

D = Distance in nautical mile,

Lat 1 = Latitude at start of haul

Lat 2 = Latitude at end of haul

Long 1 = Longitude at start of haul

Lang 2 = Longitude at end of haul

Exact latitude and longitude values can be obtained from global positioning system (GPS).

If exact positions are not available, the distance covered for an hour can be calculated from

$$D = \sqrt{VS^2+CS^2+2*VS*CS*\cos(dirV-dirC)}$$

where,

D = Distance in nm

VS = Velocity of vessel (knots = nm/hr)

CS = Velocity of current (knots)

dirV = Course of vessel (degrees)

dirC = Direction of current (degrees)

Biomass Estimation

(i) Catch in Weight per unit Area

Catch in weight per unit area can be calculated by

$$\frac{Cw/t}{a/t} \frac{Cw(kg/nm^2)t}{a} = \frac{Cw/t}{a/t}$$

where,

Cw = Catch in weight of a haul

Cw/t = Catch in weight per hour

t = Time spent for hauling (in hours)

a = Area swept in sq.km

a/t = Area swept per hour

(ii) Average Biomass per unit Area

Average biomass per unit area can be calculated by

$$\bar{b} = \overline{(Cw/a)} \times 1 Kg/nm^2$$

where,

$\bar{b}$ = Average biomass per unit area (Kg/nm^2)

$\overline{(Cw/a)}$ = mean catch per unit area of all hauls

x1 = Fraction of biomass in the effective path swept by trawl which is actually retained in the gear (usually 0.5 - 1.0)

(iii) Total Biomass for Total Area under Investigation

Total biomass could be estimated for total area under investigation by

$$B = \frac{\overline{(Cw/a)} \times A}{x1}$$

where,

B = Total biomass

A = Total size of the area under investigation (nm^2)

Though swept area method is not very precise, higher precision can be obtained by increasing the number of hauls. Another way of reducing errors is to apply stratified sampling. Suitable stratification may reduce the variance considerably for the same number of hauls and thus improve survey efficiency. The distribution of many species is determined by depth and bottom type.

Estimation of Potential Yields in (more or less) Virgin Stocks

When the available data on the stocks in certain areas are not sufficient for accurate estimation of stock, the following equation could be used for assessing potential yields (= Py $\approx$ MSY). In this chapter, few empirical formulae are given for estimate of MSY based on scanty data. These formulae could be used for bottom trawl surveys/acoustic surveys.

Methods of Estimation of Potential Yields

The methods include

- ☆ Gulland equation
- ☆ Ricker equation
- ☆ Combination of Gulland and Ricker equation
- ☆ Cadima's formula
- ☆ Schaefer and Fox model

(i) Gulland Equation

- ☆ The following equation is most commonly used for estimation of potential yields.

  MSY = M * 0.5 * Bv. (1)

  Bv is the Virgin standing stock (Estimated in a trawling or acoustic survey)

  M = Natural mortality coefficient

- ☆ For less exploited stocks, the following equation is used.

  MSY = Zt 0.5 Bt. (2)

  Zt is the exponential rate of total mortality (F + M = Z) of the year t.

  Bt = standing stock size in that year

(ii) Ricker Equation

- ☆ To estimate MSY, the equation of Ricker could be used

  $$MSY= \frac{r_m * \overline{B\infty}}{4} \quad (3)$$

  B∞= carrying capacity of the environment for a given stock (This is assumed to be corresponding to Bv, the Virgin stock size).

  'rm' = " Intrinsic rate of increase " of the population

  ['rm' could be computed using the equation of Blueweiss *et al.* (1978)]

  $rm = 0.025 * W^{-0.26}$ (4)

  rm = expressed in daily basis

  w = is the mean weight (in g.) of the adult animals under consideration.

(iii) Combining the above two equations (Ricker and Bluewiess *et al.,* 1978) and converting to the year as a time unit the following equation may be written.

  $MSY = 2.3 * W^{-0.26} * B_v$ (5)

- ☆ This equation can be used to estimate potential yields when virgin stock size and mean weight (in g) of the adult in that stock are known.

(iv) Cadima's Formula

- ☆ For estimation of exploited fish stocks for which limited stock assessment data are available, following equation is used.

  $MSY = 0.5 * (Y+M*\overline{B})$ (6)

  Y = total catch in a year

  $\overline{B}$= the average biomass in the same year.

  M = Natural mortality

☆ The above equation could be used for assessing the stock where catch and effort time series are not available and in such cases the biomass estimates could be obtained from trawl or acoustic surveys.

(v) Estimation of MSY

Using Schaefer and Fox model MSY could be estimated by following equation:

$$MSY = -0.25 \times \frac{a^2}{b} \text{ (Schaefer)}$$

$$MSY = -\left(\frac{1}{d}\right) \times \exp.(c-1) \text{ (Fox)}$$

$$FMSY = -0.5 \times \frac{a}{b} \text{ (Schaefer)}$$

$$FMSY = -\left(\frac{1}{d}\right) \text{ (Fox)}$$

In Schaefer equation, 'a' is the intercept and 'b' is the slope and this is replaced by 'c' and'd' in Fox model.

Chapter 13

Analytical Models

Future yields and stock biomass levels can be predicted by means of mathematical models. The prediction models can be used to forecast the fishery of a stock and by adopting suitable management measures pertaining to optimization of fishing effort and increase or reduction of fishing fleets, regulation of changes in mesh sizes, closed season, closed areas *etc.* The stock should be sustained without depletion. Therefore these models form a direct link between fish stock assessment and fishery resource management. This model was developed by Baranov (1914), Russel (1931), Thomson and Bell (1934) and Beverton and Holt (1956). This model requires a length and age composition of fish as input data. Further, the prediction models also incorporate aspects of prices and value of the catch, hence these models are also suitable for bio-economic analysis.

Yield/Recruit Model of Beverton and Holt

The Beverton–Holt model is a classic discrete-time population model which gives the expected number n_{t+1} (or density) of individuals in generation t + 1 as a function of the number of individuals in the previous generation.

The yield/recruit model of Beverton and Holt is also called as analytical model. This model is in a principle a "Steady state model". This means that the model describes the state of stock and the yield in a situation when the fishing pattern has been the same for such a long time that all fish alive have been exposed to it since they recruited. (Sparre and Venema, 1998).

Using a right mesh size, yield is optimized from a given number of recruits. According to Gulland (1983), calculation of yield from a given recruitment, known as yield per recruit, is a basic element in the assessment of fish stocks.

Assumption to be taken for using Analytic Model

- ☆ Recruitment, fishing and natural mortalities in a stock should be constant.
- ☆ All fish of a cohort are hatched on the same date
- ☆ Recruitment and selection are 'Knife-edge'.
- ☆ There should be complete mixing within the stock
- ☆ Fishing mortality and natural mortality should be constant during exploitation phase.
- ☆ The length-weight relationship has an exponent 3, *i.e.* $W = a \times L^3$.
- ☆ Under these assumptions, the yield from a cohort during its life span is equal to the yield from all cohorts during a year.

Derivation of the Model

At birth the cohort of a fish has age zero. The cohort grows and attains Tr, which is called as 'age at recruitment.' From 'O' age to Tr, the stock is in the pre-recruitment phase. From age Tr to Tc (age at first capture), the cohort is not experiencing any fishing mortality. In these lengths, the fish escapes through the meshes if they enter the gear. From age O to Tc, the cohort experiences only natural mortality and this is assumed to be constant through the entire life span of the cohort. At age Tc, the fish start to be caught with the mesh size actually in use and from age Tc onwards, the fish experiences fishing mortality. The fishing mortality is also assumed to be constant throughout the life span of the cohort.

Input Data Needed

1. Growth parameters of a stock
2. Mortality parameters of a stock
3. Selection parameters of a stock

Equation

$$Y/F = F \bullet \exp\left[-M^*(TC-Tr)\right]^*W\infty^*\left[1/Z-3S/Z+K+3/S^2/Z+2K-S^3/Z+3K\right]$$

where,

S = EXP. [-K* (Tc – T)]

K = Growth coefficient or curvature parameter.

T_0 = Initial condition parameter

Tc = Age at first capture

Tr = Age at recruitment

W∞ = Asymptotic body weight

F = Fishing mortality

M = Natural mortality

Z = F + M = Total mortality

Calculation Procedure

Yield/recruit is calculated for a tropical species as a function of F. This is because 'F' is proportional to effort. Using different 'F' values, an optimal 'F' value could be ascertained to give maximum sustainable yield per recruit. The optimal 'F' value is denoted as FMSY and the corresponding yield is called as Maximum Sustainable Yield. Thus by testing various F values, maximum value of Y/R, the maximum sustainable yield per recruit (MSY/R) could be achieved.

Beverton and Holt's Relative Yield per Recruit Model

Fishing effort plays a major role in all fisheries management purposes. If the 'F' is increased concomitantly, yield will decrease. Under this circumstance, calculation of Y/R in grams per recruit will not serve the purpose. Hence, Beverton and Holt developed a model as Beverton and Holt Relative yield per recruit model (Y/R)'. This model will serve as good tool for providing information needed for management. With the help of few parameters and optimal prediction of mesh size regulations, relative yield per recruit could be calculated. Further, this model requires only length of fish rather than ages.

Equation

The relative yield per recruit is defined by

$$(Y/R)' = E * U^{M/K} * \left[1 - \frac{3U}{1+M} + \frac{3U^2}{1+2M} + \frac{3U^3}{1+3M}\right]$$

where,

$$M = \frac{1-E}{M/K} = K/Z$$

$$U = 1 - LC/L\infty$$

$$E = F/Z$$

The (Y/R)' is mainly attributed to U and E. The needed parameters are M/K. The values (Y/R)' could be plotted with values of E ranging from 0 to 1, with corresponding F values from 0 to ∞. A curve thus obtained gives a maximum value of EMSY for a given value of Lc. Thus knowing Lc, L∞ and M/K for a certain fishery, the E exploitation rate can be compared with EMSY level. Accordingly, better management strategies could be proposed.

Yield per Recruit from Length Data

When L∞ and Lr are known instead of t_0, Tr and Tc, the Yield per recruit (Y/R) from length data could be obtained from

$$Y/R = F * A * W\infty\left[1 - \frac{1}{Z} - \frac{3U}{Z+K} - \frac{3U^2}{Z+2K} - \frac{3U^3}{Z+3K}\right]$$

where,

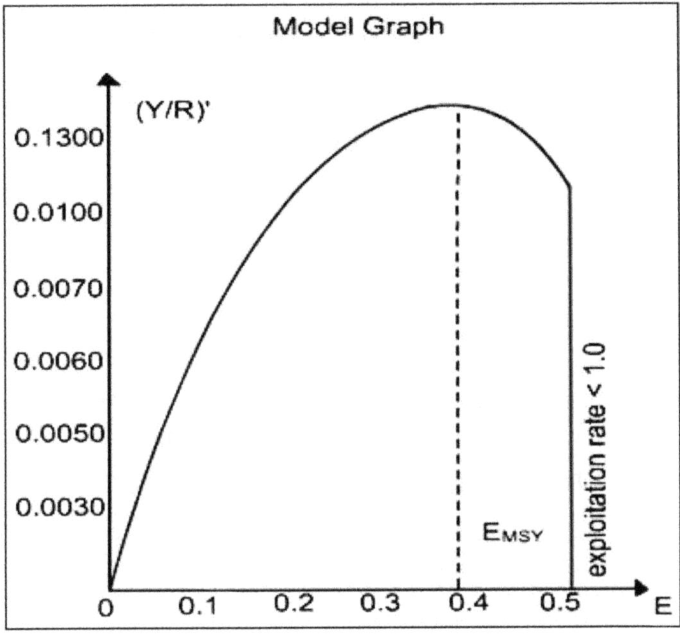

Figure 29

$U = 1 - L_c/L\infty$

$A = [L\infty - L_c/L\infty - L_r]^{M/K}$

Biomass per Recruit Model

The model of Beverton and Holt, yield per recruit model could be used to determine annual average biomass of survivors as a function of F. The average biomass is related to CPUE.

The equation is

$$YR = F \times \overline{B}/R \qquad (1)$$

The formula used to calculate B/R is the same as in equation - 1 divided by the F. The average biomass, B/R thus obtained is considered to be the biomass of exploited part of the cohort *i.e.* the biomass of fish of age T_c or older.

Mean Age and Size in the Yield

When Z is constant from time of recruitment (Tr) and age at first capture (Tc), mean age and mean length in the annual yield could be estimated with the following equation.

$$\overline{Ty} = \frac{1}{Z+T_c}$$

Mean length in annual yield, where $\overline{Ty}$ is the mean age in yield

Similarly mean length in the annual yield is $\overline{Ly}$ = L∞ [1-Z* S]/Z+K∞

S = exp [-K * (T_c – T_0)] = 1 – L_c/L∞

Tc or Lc can be replaced by any age from which the fish have a constant mortality, so as to give mean length in that part of population.

Mean Weight in Annual Yield

$\overline{Wy}$ = Z * W∞ [1/Z + 3*S/Z+K + 3* S^2/Z+2K + S^3/Z+3*K]

Whereas the three parameters $\overline{Ty}$, $\overline{Ly}$, $\overline{Wy}$ and exploited biomass along with CPUE will decrease by increasing Z (*i.e.*) with effort.

In the unexploited fishery, the decrease may be faster for low values of F. In all the three parameters ($\overline{Ty}$, $\overline{Ly}$, $\overline{Wy}$), T_c forms a common input as Tc is determined by mesh size. When the mesh size is large, the mean age and size will be higher.

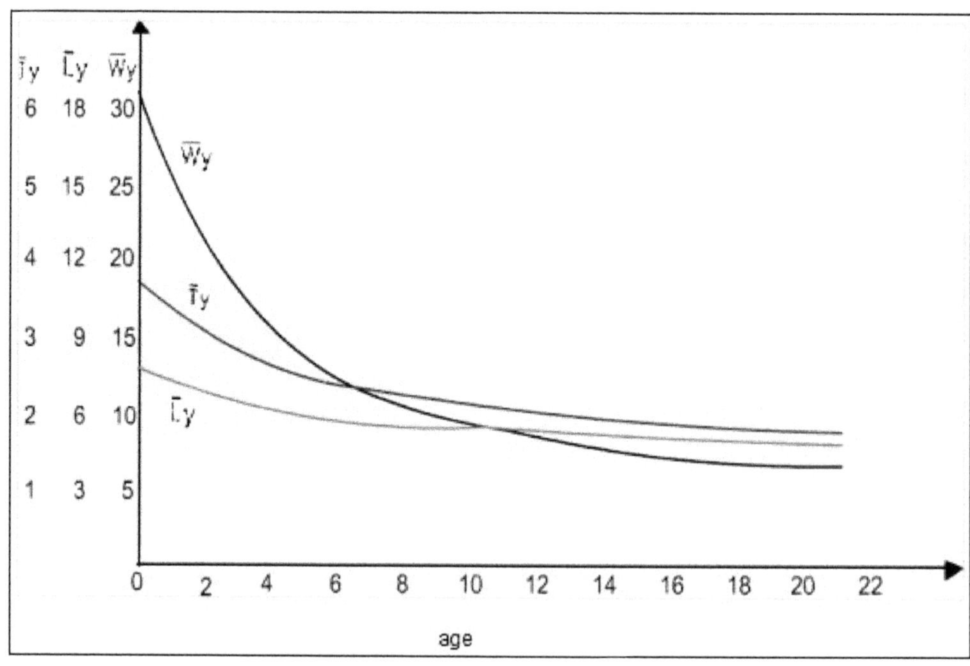

Figure 30

Yield Curves

The information needed to draw yield curves with usage of analytic models is growth rate and natural mortality of fish.

'S' shaped curve results growth in weight over time.

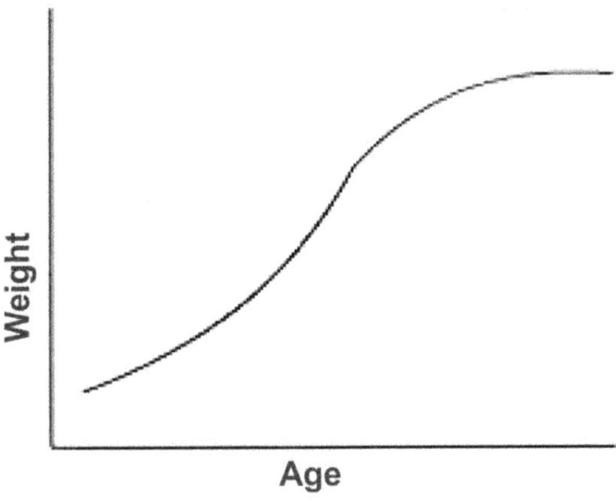

Figure 31

An exponential decline occurs with survival over time.

From the biomass recruit model, biomass per recruit curve will be obtained. This curve will always decrease by increasing the effort. In any fishery with decreasing in CPUE, the biomass will increase when effort increases.

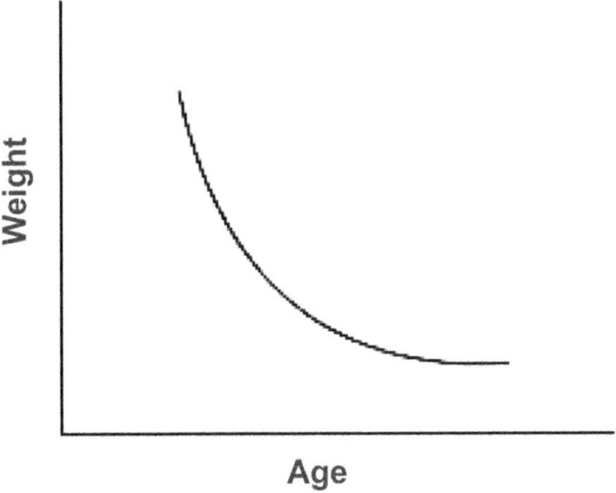

Figure 32

From these two curves, it is possible to derive yield per recruit curve for a fixed age of fish capture T_c. The T_c is the age at which the fish becomes vulnerable.

In yield/recruit curve, F is taken as independent variable and Y/R is dependent variable. Knowing the yield, keeping T_c constant, for a given value of F, the number of recruits could be ascertained by dividing the total yield by yield in gms of recruit.

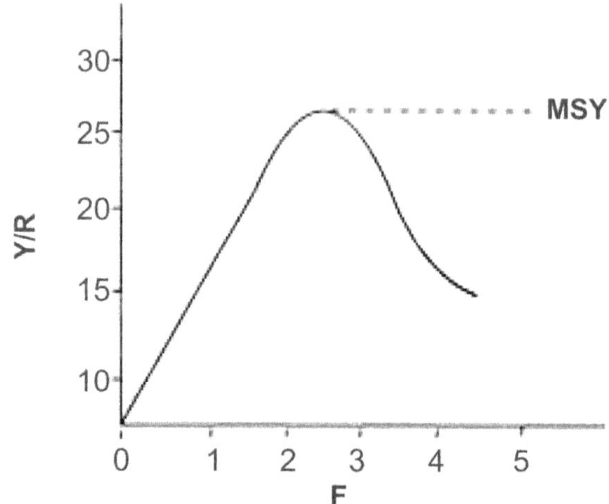

Figure 33: Yield Assessment with Reference to Various Level of Efforts

The curve drawn in the above Figure 33 is called as yield per recruit curve, and the peak of the curve is MSY.

Yield isopleths could be drawn for various combinations of F and T_c.

The yield per recruit curve depicts the MSY. The MSY depends on age at first capture T_c, and in turn T_c depends on mesh size used for a fishery.

The following curves show how the yield is affected with T_c.

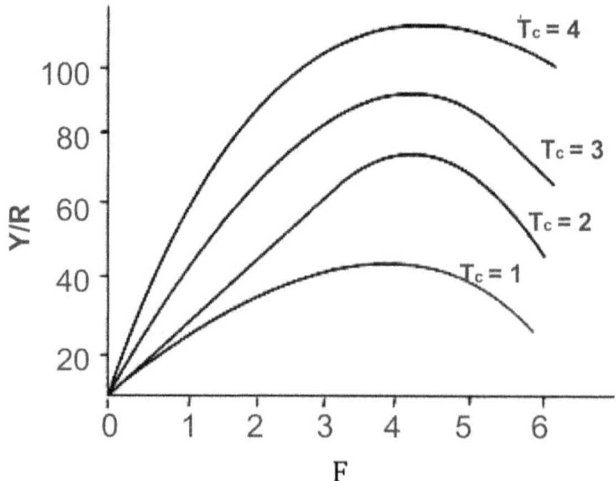

Figure 34: Yield per Recruit Variation Due to Change in Age at First Capture

MSY is highest at highest value of T_c with higher value of fishing effort.

The T_c and F could be managed by stock assessment scientist/fishery mangers to have highest MSY.

'F' is proportional to effort.

T_c is the function of gear selectivity.

Thus combining a range of values of T_c with a range of values of F, sustainable yield could be achieved, for a certain level of T_c and F.

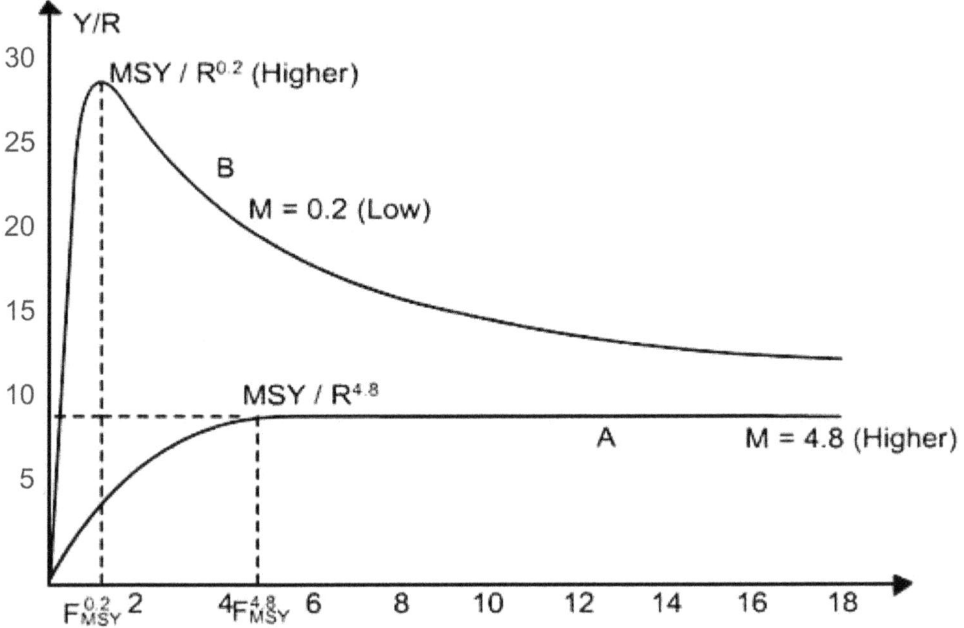

Figure 35: Yield per Recruit Curve as a Function of F

In the above graph, curve B is higher compared to curve A. But it has lower value of FMSY but with higher MSY/R. The main difference between two curves is the natural motality rate 0.2 for the curve B and 4.8 for the curve A. Several parameters can influence yield curve. Among them 'M' affects the yield curve. The variations in M can influence on the shape of the curves Y/R. Thus when M is lower, lower FMSY is produced and a higher MSY/R. Increasing the fishing effort above FMSY leads to decrease in total yield. When M is high, FMSY is difficult to estimate in the Y/R curve.

Chapter 14

Computer Packages for Stock Assessment Studies

Computer applications are widely used in fishery science particularly in fish stock assessment and fish taxonomical studies. Fisheries science was slow to adopt computers on a wide scale basis. Since computer made easy all the calculations, its finds wide application in stock assessment studies. They are not only solving biological problems but also give many simulations for the fishery managers.

Several packages were developed for stock assessment studies in the temperate waters. Packages were also developed exclusively for tropical waters by FAO and ICLARM during the year 1987 and 1988 respectively. Computer packages for tropical waters are mainly based on the length based data.

1. ELEFAN Packages

The ELEFAN system (Electronic Length Frequency Analysis) was developed by ICLARM in response to the need for robust methods for analysis of length frequency data and the availability of cheap microcomputers. This system consists of five programmes.

1. ELEFAN 0
2. ELEFAN I
3. ELEFAN II
4. ELEFAN III
5. ELEFAN IV

ELEFAN O

This is basic system used to create and modify length frequency data files for the use of other four programmes. ELEFAN I to IV will be using length frequency data created by ELEFAN-O

ELEFAN-I

It was developed by Pauly and David (1987). This programme is used to estimate the growth parameters of fish or invertebrates. Growth oscillations of fish can be correlated with selected environmental parameters.

ELEFAN-II

It performs a variety of computation of which the following are the main ones.

1. Estimation of total mortality (Z) and derived quantities from the straight, descending arm of a length converted catch curve.
2. Estimation of probabilities of capture by length and mean length at first capture (Lc) from the ascending left arm of a length converted catch curve.
3. Expressions of seasonal changes in recruitment intensity in the form of graphical recruitment pattern.

It requires no inputs other than growth parameter estimates and length frequency data.

ELEFAN-III

It incorporates 3 types of Virtual Population Analysis (VPA).

VPA – I

It estimates standing stock (in numbers) and fishing mortalities by time intervals.

VPA – II

This programme is used to estimate standing stock (in number) and fishing mortalities by length class in a stock with age distribution.

VPA – III

It provides estimates of standing stock and fishing mortalities by month by length. Its requires monthly catch data in addition to length frequency and growth parameters as input.

ELEFAN IV

This programme provides gear selection parameters (Tc the probability of capture by length class) and can be used to estimate or and probabilities of recruitment by length class from catch samples.

2. LFSA Packages

The Packages, Length – based Fish Stock Assessment (LFSA) was developed by Sparre during 1987. This is suitable package for ordinary IBN Computers. It requires only length frequency data as input. This package was specially designed for tropical waters divided into four sub-packages according to the type of input data.

1. LF – Length frequency analysis
2. Al – Age/Length data
3. Beverton and Holt Y/R model
4. Ordinary linear regression

This package is provided with 32 sub-packages, and are available to calculate various growth parameters, mortality parameters, analysis, Y/R model *etc.* its requires length frequency, effort, total catch, position in terms of latitude and longitude, depth, gear type as inputs.

3. FiSAT

FiSAT is relatively new package and the acronym stands for FAO-ICLARM Stock Assessment Tools. This software is resulted from the merging of its two predecessors, the complete ELEFAN package developed by ICLARM and LFSA developed by FAO with addition to new routines found useful for the analysis of length frequency. FiSAT was developed during the year 1995 by Gayanilo, Sparre and Pauly. FiSAT wad developed for an IBN or compatible personal computers with a minimum of 512 KB of free disc space and primarily aimed at helping scientists working on tropical fisheries resources to produce scientific advice for fisheries management.

This length assessment tool was developed to analyse L/F data -at age, catch – at- age, selection and other data typically collected for tropical fish stock assessment. Its takes advantage of the high resolution graphic capabilities of micro computer.

FiSAT has four main groups or routines.

1. File
2. Assess
3. Support
4. Utilities

File

It deals with the file creation, editing and various data manipulation.

Assess

It contains the models and methodologies used in the analysis of the different types of data that FiSAT supports, ranging from growth and mortality parameters estimation to prediction using either the Beverton and Holt Y/R analysis, or the Thompson and Bell yield and stock prediction models.

Support

This routine facilitates data analysis. It contains routines to simulate length frequencies, display bar diagram estimate (Unsampled) maximum lengths and sample weight and perform regression analysis.

Utilities

It contains functions for importing and exporting data files, configuration output options and managing some DOS functions and a simple calculator is also included.

4. VIT

Software for fishery analysis published in 1998 by FAO, and is created for the analysis of fisheries where the information is limited. The package is designed to analyse exploited marine populations based on catch data, structured by age or size and by type of gear.

5. POPDYN

A population dynamics database allows to user the store stock based information on the characteristic of fisheries system. It was developed by FAO in 1994.

6. SPATIAL

It was developed by FAO in 1994. A simulation package developed to model the space/time distribution of fishing intensity using alternative approaches.

7. ANALEN

It was developed by Chevaillier and Laurec, 1990 in French for FAO in 1990. It is a software for analysis of catch at length data for simulation of multi gear fisheries.

This package contains four programmes.

1. ANALEN- based on Jones length based cohort analysis.
2. ANAJON Calculates Y/R and mature biomass per recruit for different levels of exploitation also provides short term and long term forecasts of catches.
3. SENJON Sensitivity analysis of the principal results.
4. MONOJO to evaluate technical interaction between fishing gear.

8. ANACO

Developed by Mensil, 1989 for FAO. This package contains two programmes related to VPA.

I. SIMUCO

It was developed to analysis the selected properties of catch at age data.

II. VPBAS-COHORT-VPUTIL

It is a software for assessment by using time series of catch at age data.

9. NAN-SIS Package

This FAO package was developed by Stromme (1992). This is specially designed to process data generated by surveys of Dr. NANSEN by swept area method. NAN-SIS is a Survey Information System for logging, editing and analysis of scientific trawl survey data (Catch and I/f data).

10. The Ecopath with Ecosim (EwE) Approach

Ecopath with Ecosim (EwE) is a free ecological/ecosystem modelling software suite for personal computers that has built and extended on for almost twenty years, initially started at NOAA by Jeffery Polovina, but has since primarily been developed at the UBC Fisheries Centre of the University of British Columbia. In 2007, it was named as one of the ten biggest scientific breakthroughs in NOAA's 200-year history. The NOAA citation states that Ecopath "revolutionized scientists' ability worldwide to understand complex marine ecosystems"

EwE has three main components:

1. Ecopath – a static, mass balanced snapshot of the system
2. Ecosim – a time dynamic simulation module for policy exploitation
3. Ecospace -a spatial and temporal dynamic module primarily designed for exploring impact and placement of protected areas.

The Ecopath software package can be used to:

- ☆ Address ecological questions
- ☆ Evaluate ecosystem effects of fishing
- ☆ Explore management policy options
- ☆ Evaluate impact and placement of marine protected areas
- ☆ Evaluate effect of environmental changes
- ☆ Analyze impact and placement of marine protected areas
- ☆ Predict movement and accumulation of contaminants and tracers (Ecotracer)
- ☆ Model effect of environmental changes
- ☆ Facilitate end-to-end model construction.

The desktop version of Ecopath with Ecosim runs only on Windows and requires Microsoft Access database driver's version 2007 or newer. The Computational core of Ecopath with Ecosim can be executed on other operating system such as Unix or Linux using the Mono common language runtime. Development of Ecopath version 6 received support from the Lenfest Ocean Program and the Pew Charitable trusts. In 2011, the Ecopath Research and Development consortium was founded to share the responsibility of maintaining and further developing the approach with institutions around the world.

11. Monte Carlo Simulation Model

The Monte Carlo method was evolved during 1940s by John von Neumann, Stanislaw Ulam and Nicholas Metropolis, while they were working on nuclear weapon projects (Manhattan Project) in the Los Alamos National Laboratory. Monte Carlo methods (or Monte Carlo experiments) are a class of computational algorithms that rely on repeated random sampling to compute their results. Monte Carlo methods are often used in computer simulations of physical and mathematical systems. These methods are most suited to calculation by a computer and tend to be used when it is infeasible to compute an exact result with a deterministic algorithm. This method is also used to complement the theoretical derivations.

Applications

Monte Carlo methods are especially useful for simulating systems with many coupled degrees of freedom, such as fluids, disordered materials, strongly coupled solids, and cellular structures. They are used to model phenomena with significant uncertainty in inputs, such as the calculation of risk in business.

Chapter 15

Stock Recruitment Relationship

Recruitment over fishing results in when the parent stock is reduced to the extent that enough young fish are not produced to compensate the stock. There has been a great deal of effort devoted to determine relationship between the number of adults and the number of recruits they produce.

Recruitment

- ✰ The process by which a fish or shrimp of an age group integrates itself for the first time in to the exploitable stock is known as recruitment, which is a function of age and is a progressive event. The animals added on to the population is called the recruits.
- ✰ The concept of stock recruitment relationship is related to the concept of surplus production. It is a general concept their when population becomes smaller, the reproductive success increases. The following are the mechanisms by which the reproductive success of a fish can increase.
  - i) Increased growth and better fecundity of a species.
  - ii) Good food supply/individual
  - iii) Good survival of eggs and larvae due to reduced cannibalision.
  - iv) Optimal fishing effort.
- ✰ The recruitment pattern is affected by environment variables. When the environment variables are at optimal level, relationship between stock and recruitment will be accurate.

Normally for growth related studies the following three models are used

1. Ricker bell shaped curve model

2. Beverton and Holt model
3. Cushing model

Ricker Bell Shaped Curve

In Ricker's model it implies that if a stock becomes very large number of recruits, are reduced. The plotting of parent stock and recruits is more bell shaped. Further this model emphasize that a relationship between the parent stock and recruitment stock exists for all size of the spawning stock and that there is an optimum spawners stock size.

The formula used to establish relationship between parent stock and recruit stock.

- ☆ $R=Pr^{a(1-P/Pr)}$ (1)

  R = number of recruits

  P = number of parents

  Pr = Replacement of parent stock *i.e.* where P=R.

- ☆ $\log_e^{a(1-P/Pr)} = R/P$ (2)
- ☆ $\log_a^{(1-P/Pr)} = \log_e(R/P) = \log_e R - \log_e P$ (3)
- ☆ $\log_e R - \log_e P = a - (a/Pr)*P$ (4)

  The above equation of linear form with a negative relationship of the type

- ☆ $Y= a - b X_1$; $Y = \log_e^{R - \log e^P}$; $X_1 = P$

Beverton and Holt Model

Beverton and Holt model suggests that when the population of parents becomes large, the increase in recruitment per unit increase in stock size becomes progressively smaller forming a sigmoid curve.

- ☆ R = (E/E+g* R max)* Rmax

  R = number of recruits

  E = number of females * average egg production.

  G = parameter where the number of recruits, R, increases towards asymptotic level, Rmax.

Cushing Model

- ☆ A linear curve is generated when plot's made between parents stock and recruits.
- ☆ The main bottle neck in this relationship is that environmental variables often appears to have a significant impact on recruitment success in a number of years. Thus over a wide range of stock densities, the amount of recruitment will be loosely dependent on stock size.

Relative Response Model

This model was developed by Alagaraja (1984). The aim of this model is that removal of more fish in one year leads to better survival and growth in the residual stock. Increasing the fishing effort leads to increase in catch and when fishing effort goes beyond maximum, the response of the stock continues to be positive but the stock will show sign of depletion. Finally this results to a nil stage (Zero level), negative response of the stock to the fishing effort.

- ☆ Ct+1= a+bCt

  Ct+1 = catches at successive periods. (Say, years)

  Ct catch in a particular year.

  The curve is sigmoid type where Ct reaches the asymptotic maximum. The maximum harvestable catch is a/1-b.

Chapter 16

Fisheries Management

Overfishing

- Overfishing is generally defined as action of exerting a fishing pressure (fishing intensity) beyond the agreed optimum level. According to Pauly (1983) overfishing is indeed primordial sin, bankruptcy of fishing management.
- Overfishing may occur as Growth overfishing. Recruitment overfishing and ecosystem overfishing.

Growth Overfishing

- Growth overfishing occurs when too many small fish are being harvested by excessive effort and poor selectivity. The young fish that became available to the fishery are caught before they can be grown to the harvestable sizes. Right selectivity *i.e.* optimum mesh size should be used so as to allow the smaller size to escape. This will allow the smaller ones to grow to attain maturity.

Recruitment Overfishing

- This is a situation in which the parent stock is reduced by fishing to the extent that not recruits are produced to ensure that the stock will maintain itself. To avoid recruitment overfishing, the young ones (the recruits) are allowed to grow and they should attain the maturity and at least they should reproduce once or two times in its life cycle. The parents should not be caught for which mesh size should be optimized to allow the spawners to escape from the net. The prolonged recruitment overfishing can also lead to stock collapse. These also occurs under unfavourable environment conditions.

- ☆ The stock-recruitment relationship is worked by three methods. Stock-recruitment relationships may appear good for temperate spawner in which the spawning in synchronized. For most of the tropical species, where the spawning season is extended and species are prolific breeders, the relationship between the stock and requirement appears not consistent.

Ecosystem Overfishing

- ☆ This is a type of overfishing which occurs by the competition and predation between taxa. This type of overfishing occurs in a mixed fishery. The ecosystem overfishing would be transformation of a relative mature, efficient system into an immature, inefficient system.

Regulatory Measures

- ☆ To avoid overfishing, the fishing effort and optimal use of mesh size are to be monitored.
- ☆ To regulate fishing effort by all year at optimum level could be accomplished by
  - (i) Limiting entry and restricting be number of vessels in the fishery
  - (ii) Limiting the quantity caught in one period of time *i.e.* quotas and
  - (iii) Prohibiting fishing in certain areas and or in certain seasons.
  - (iv) Control of age or size at first capture. Once this is accomplished the overfishing of stock can be avoided and the stock could be sustained.

Eumetric Fishing

- ☆ This scheme calls for gear restrictions to achieve a right age composition of the catch function for a given level of effort. The demerits of eumetric fishing is that it forces the fleet on to a higher cost curve dissipating thereby the potential economic benefits. Further there will not be any economic gains by allowing fish to grow to a size that in eumetric with gear because the marginal revenue from growth is affect by marginal cost of programme implementation and also by M. *i.e.* natural mortality.

CPUE

An understanding of fishing effort is fundamental to understanding the assessment and management of fish stocks. Effective management involves deciding directly or indirectly upon the amount of effort which should be applied to the stock and therefore requires a proper measure of effort. In general, the effort put by a fisher for fishing should conceptually be defined as fishing effort.

Fishing Effort

Fishing effort is the product of the amount of gear in use and the duration of the fishing activity. *e.g.* trawling hours, pot or trap days, diving hours, hook hours, *etc.* Clearly each type of gear (trawl, seine, trap, *etc.*) catches fishes differently. Even within a given gear type, there can be enormous variability. For example, there may be hundreds of different types of traps or nets in use in a particular area. Furthermore, not all fishermen are equally skilled. Skills change continuously, and improvements in gear are constantly being introduced.

For example, if 15 boats, each fish for 10 days, the fishing effort is 150 days.

Catchability Coefficient

For a fishery manager to estimate the amount of fishing effort, it is not sufficient to count the amount of gear. One must count or estimate the amount of gear-days (or other measure) for each type of gear and derive empirical correction factors to express all the fishing effort in terms of one (or a few) standardized fishing gears. The task is indeed formidable. Careful gear comparisons are an important part of fishery management.

The fishing mortality, F, is almost assumed to be proportional to fishing effort. Mathematically, this is expressed as

$$qf = F$$

Where f is the amount of fishing effort (*e.g* in boat-days) and q is the proportionality coefficient, frequently referred to as the catchability coefficient. The more the efficient the gear is, the higher the value of 'q' because 'q', is the measure of the ability of the gears to catch the fish.

In the estimation of mortality parameters in surplus production models and in prediction models, the catch and effort statistics are taken as an input data. According to Rothschid (1970), the current estimation of the quantity of fishing effort is not only invaluable in following changes in abundance through CPUE index, but fishing effort is also used in (i) the relation between total catch and effort (ii) the inter-relation between yield per recruit and size of the fish caught (particularly the smallest fish caught) and fishing effort (iii) the measurement of the relation between stock and recruitment.

Standardization of Fishing Effort

In tropical fisheries, different gears are used to capture the same resources. For instance, the sharks are captured by gillnets, trawls, hook and lines, *etc.* The catching efficiency between these gears differs widely. It is also difficult to compare effort of different vessels operating a single gear. For example, horsepower of trawler engines ranges from 80 to 150, and hence, fishing hours multiplied by horsepower may be a suitable measure of effort in trawl fisheries. In a gillnet fishery, the engine horsepower and fishing hours are less important compared to the number of gillnets set per day. In a hook and line fishery, it may be appropriate to consider the number of fishermen multiplied by the number of hooks used.

An attempt to combine the effort of these gears encounters intricate problems. Sparre and Venema (1992) suggested a method in which the quantities of yield and CPUE are proportional to effort.

(i) Relative effort

$$\frac{\text{Yield}}{\text{CPUE}} = \text{Effort or CPUE} = \frac{\text{Yield}}{\text{Effort}}$$

(ii) Relative CPUE

The effort of a particular gear (gear i) in year is defined as relative catch per unit effort

$$Ri(y) = \frac{\text{CPUE}_i(Y)}{\overline{\text{CPUE}_i}(Y_1, Y_2, \ldots\ldots)} \quad (1)$$

The normalized relative effort E(y) for the particular years is calculated by

$$E = \frac{Y\,T(y)/R(y)}{\text{mean } y\,T/R}$$

Open Access Fishery

The open access fishery is one where there is no restriction placed upon utilization of the resource. When property rights for natural resource are not enforced, overexploitation of the resources occurs frequently. Hence, in open access fishery, the resource is highly exploited down to the level at which marginal profit equals zero that is no harvest beyond the highly exploited level.

Open Access Catch

At open access catch (OAC), the cost equals revenue. However, there will be normal return on capital; the fishery is characterized by over-capitalization, overcapacity, low return on investment and dissipation of rents that would have accrued to the industry (Devaraj, 1983). However, consumer's surplus will be at its maximum.

The Maximum Economic Yield (MEY) is less than OAC or MSY. However, the surplus production will be maximum and the boat owners will be making profit and the consumer surplus will be reduced as shown p, a, b, in the Figure 36.

The most important characteristic of the open access fishery is that profits are zero. Though the characteristic of open access is termed as tragedy of commons, the open access and common property are not the same. Common property of resources, tend to display restrictions on who may exploit the resource and also the execution of the exploiting resource. Open access will be a good representation of the exploitation of a resource for a certain time period.

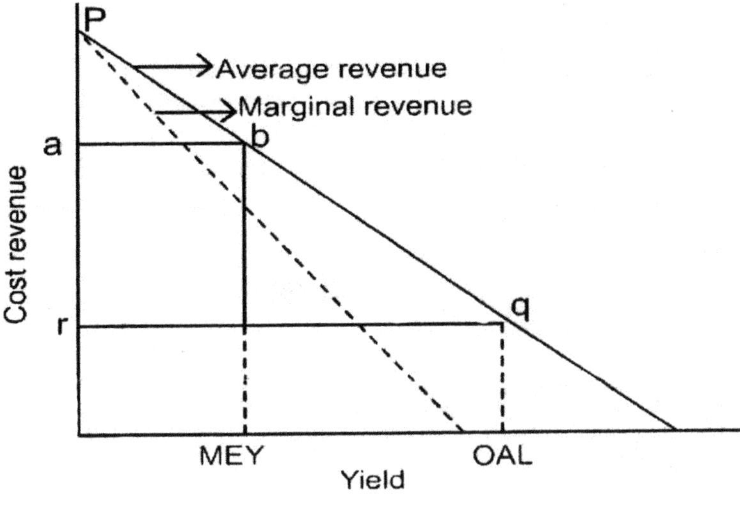

Figure 36

The good example is whale fishing. Whale fishing has potential effects of open access and at the same time by recognising that resource at global level becomes common property.

The supply-demand curve at MEY, SaY, OAC and MSY is given in Figure 37.

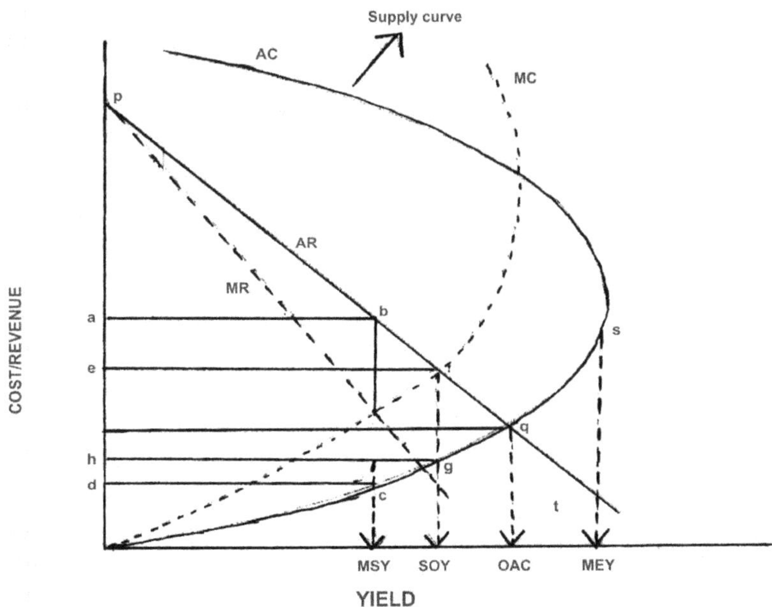

Figure 37

i) In Open Access Catch (OAC), consumer's surplus will be at its maximum (Shown in p.q. and r I the Figure 37)

ii) In MSY, he cost of fishing at a point 'p' for MSY is higher than the corresponding revenue (at t) and also the cost at open access (q). In MSY, the advantage is that the stock will be fished at its maximum sustainable level and there will be a high enough catch in the consumer with a super normal cost to the industry.

If at all, MSY is needed to be a policy, the demand curve should shift upwards intersections the cost curve at its backward bending portion so that the cost of MSY is less that at open access.

In Socially Optimal Catch, also called Socially Optimal Yield (SOE), equitable distribution of rent is achieved by reducing the private profit to the area e,f, g, h and by increasing the consumers surplus to p,e,f at the level of MEY.

Methods of Regulation

The question of management essentially is concerned with the means of rationalization of an open access fishery. Rationalization schemes popularly followed in fisheries management are: direct limitation of output, eumetric fishing, taxation and direct limitation of input through licensing. The merits and demerits of these schemes are dealt with below.

1. **Direct limitation of output**: This scheme is exemplified by the traditional closed seasons and quota system followed in many developed countries. They place a direct limitation on the output through individual quotas. Nevertheless, the scheme is disadvantageous since it is very difficult to enforce and also creates a disruption in supplies.
2. **Eumetric fishing**: This scheme calls for gear restrictions to achieve a right age composition of the catch for a given level of effort. The demerit of this scheme is that it forces the fleet on to a higher cost curve dissipating thereby the potential economic benefits. No economic gains can be obtained by allowing the fish to grow to a size that is eumetric with the gear since the marginal revenue from growth is just offset by: (a) the marginal cost of programme implementation, and (b) the marginal losses to natural mortality (Crutchfield, 1979).
3. **Taxation**: Taxation plays a dual role in the rationalization of a overcapitalized fishery. One of the uses of taxation is that it can act as a means of offsetting the effects of an otherwise efficient management regime (*e.g.,* licence limitation programme) on the distribution of wealth, income and employment. A highly successful limited entry programme unaccompanied by any tax measure to capture a part of the economic rent created may effect transfer of income and wealth in a direction unacceptable to the society. Any measure that confers substantial gains on

private enterprises at considerable social costs is not favoured. The major advantage of a tax on the remaining participants in an efficient fishery is its ability to convert the social costs of management to an explicit charge on the productive activity that gives rise to them. The other use of taxation is its ability to influence the level and composition of effort; it can be so devised as to convert the several types of externality plagueing an open access fishery into contractual costs to the individual decision maker.

4. **Direct limitation of input**: This scheme may be implemented through a licence limitation programme which is preferred to all other schemes, for, it is easily enforcible and causes no disruption in supplies. Yet, it is confronted with certain important issues as: (a) the controversy over licensing the vessels or fishermen; (b) allocation of licences between different and conflicting interests (*e.g.,* different gear or vessel types operating on a single stock); (c) number of licences to be made available; and (d) criteria for basing the issue of licences.

International Plans of Actions (IPOAs)

International Plans of Action (IPOAs) have been developed by FAO in order to facilitate effective implementation of Code of Conduct for Responsible Fisheries (CCRF). These are voluntary instruments elaborated within the framework of the CCRF and apply to all states and stakeholders. IPOAs developed so far pertains to the following areas:

1. Reducing incidental catch of seabirds in long line fisheries (FAO, 1999a)
2. Conservation and management of sharks (FAO, 1999a)
3. Management of fishing capacity (FAO, 1999a) and
4. Prevention of illegal, unreported and unregulated (IUU) fishing (FAO, 2001).

Bycatch Reduction Technologies

'Target catch' is the species or species assemblage primarily sought in a fishery, 'incidental catch' is the retained catch of non-targeted species and 'discarded catch' is that portion of catch returned to the sea because of economic, legal or personal considerations (Alverson *et al.,* 1994). Bycatch includes both discarded and incidental catch. In addition to the non-targeted finfishes and invertebrates, bycatch also involves threatened and protected species like sea turtles.

Different types of bycatch reduction technologies have been developed in the fishing industry around the world. Devices developed to exclude the endangered species like turtle and to reduce the non-targeted species in shrimp trawling are collectively known an Bycatch Reduction Devices (BRDs). These devices have been developed taking into consideration variation in the size and differential behaviour pattern of shrimp and other animals inside the net. BRDs can be broadly classified into three categories based on the type of materials used for their construction, *viz.,* Soft BRDs, Hard BRDs, and Combination BRDs. Soft BRDs make use of soft materials like netting and rope frames for separating and excluding bycatch. Hard BRDs are those, which use hard or semi-flexible grids and structures for separating and excluding bycatch. Combination BRDs use more than one BRD, usually hard BRD in combination with soft BRD, integrated to a single system. Juvenile Fish Excluder cum Shrimp Sorting Device (JFE-SSD) is a Smart Gear award winning design (WWF) developed by CIFT for protecting juveniles and for pre-sorting of the catch.

Pointers from CCRF and IPOAs for Responsible Fishing

FAO Code of Conduct for Responsible Fisheries provides the following pointers for responsible fishing and sustainable fisheries development:

- ☆ Evolve regionalized consensus of Code of Conduct for Responsible Fishing, in close participation with all stakeholders (traditional, motorized and mechanized fishermen organizations), fisheries research organizations and fisheries managers.
- ☆ Maintain a registry of all fishing vessels in waters under State jurisdiction with all essential details.
- ☆ Take measures to control open access by strict enforcement of a system of licenses (authorization to fish) in traditional, motorized and mechanized sectors.
- ☆ Periodically revalidate maximum sustainable yield of resources in the existing fishing grounds and determine fishing units of specific capacity in each category, for sustainable harvesting of resources.
- ☆ Standardise the capacities, dimensions and specifications of fishing units in each category.
- ☆ Address the question of excess capacity and take steps to remove excess capacity over a time schedule.
- ☆ Identify and delimit Protected Areas in marine and inland water ecosystems.
- ☆ Conduct periodic audit of fishing craft and gear combinations, their economics of operation, ecological and environmental impacts.
- ☆ Evolve regulations for mandatory survey of mechanized fishing vessels.
- ☆ Evolve a system for marking fishing vessels and fishing gears.
- ☆ Evolve regulations and promote use of life saving, fire fighting and communication equipment for safety of fishermen.
- ☆ Evolve regulations for mandatory survey of mechanized fishing vessels.
- ☆ Promote selective fishing gear and practices.
- ☆ Develop and implement National Plans of Action (NPOAs) for (i) management of fishing capacity, (ii) prevention of illegal, unreported and

unregulated (IUU) fishing, (iii) conservation and management of sharks, and (iv) reducing incidental catch of seabirds in long line fisheries.

- Evolve an efficient Monitoring, Control and Surveillance (MCS) system.
- Make effective use of Geographical Information System for fisheries management; monitoring and control of fishing effort and energy use.
- Evolve and promote a package of practices of energy conservation in fish harvesting.
- Develop a Fisheries Information System for providing easy access to authentic information and facilitating fisheries research, management and business.
- Evolve a mandatory programme of training and certification for non-motorised, motorized and mechanized fishermen in safe navigation, fisheries regulations, responsible fishing, log keeping and reporting

Implementation of the New Law of the Sea

All coastal countries now have a recognized right to use and regulate the fishing in their coastal zones. The new Law of the Sea created a massive reallocation among nations of the right to fish. The 200-mi zone along coasts and around islands contains about 99 per cent of the world's land. Much of the catches from these stocks had been taken by the distant-water fishing fleets of about two dozen countries that traditionally had fished in many areas without restraint and without even maintaining the records of catch and effort that are essential for tracking the condition of the stocks. Particularly damaging had been the practice of "pulse" fishing, in which some fleets fished an area intensively until the fishing became unprofitable and then moved to another area.

Now coastal countries have the resources allocated to them. They have the opportunities to obtain optimum yield, to allow their citizens to fish the coastal waters, to lease any part of the fishing rights to foreign fishermen, and to use access to the resources as part of a bargain, such as a joint venture in which the country receives a fish processing facility and part of the catch in return for the fishing rights.

Along with the opportunities come obligations. The law requires control of fishing to ensure conservation, which requires scientific knowledge of the resources and statistical information on the fishing. The law also requires granting of access to foreign fishermen if the coastal state does not harvest the optimum catch.

Taking advantage of the opportunities and complying with the obligations will require most countries to form new or enlarged fishery regulatory and development organizations. Regulatory organizations will need competence to perform research, collect statistics on the fishing, negotiate with foreign countries about fishing and fishing boundary zones, make decisions about regulations, and enforce regulations. Development organizations will need competence to perform economic, social, and organizational planning, as well as knowledge of the business of fishing processing, and marketing. All countries will have boundary problems with their neighbours and will probably join regional fishery organizations. Implementation will be a long and continuing process.

Excerpts Pertaining to Fisheries from the Law of the Sea

After centuries of piecemeal development of the Law of the Sea, a process began in 1969 under the United Nations to prepare a comprehensive legal code. It was agreed to cover 25 main subjects, most of which involve several issues.

The subjects pertaining especially to fisheries included:

- Continental shelf;
- Exclusive Economic Zone beyond the territorial sea;
- Coastal state preferential rights or other nonexclusive jurisdiction over resources beyond the territorial sea;
- High seas;
- Landlocked countries;
- Rights and interests of shelf-locked states and states with narrow shelves or short coastlines;
- Preservation of the marine environment;
- Scientific research.

Conservation of the Living Resources

- The coastal State shall determine the allowable catch of the living resource in its exclusive economic zone.
- The coastal State. Shall take into account the best scientific evidence available to it. Shall ensure through proper conservation and management measures that the maintenance of the living resources in the exclusive economic zone is not endangered by over-exploitation. As regional, regional or global, shall cooperate to this end.
- Such measures shall also be designed to maintain or restore populations of harvested species at levels which can produce the maximum sustainable yield. As qualified by relevant environmental and economic factors, including the economic needs of coastal fishing communities and the special requirements of developing States, and taking into account fishing patterns. The interdependence of stocks and any generally recommended international minimum standards whether subregional, regional or global.
- In taking such measures the coastal State shall take into consideration the effects on species associated with or dependent upon harvested species above levels at which their reproduction may become seriously threatened.

- Available scientific information, catch and fishing effort statistics, and other data relevant to the conservation of fish stocks shall be contributed and exchanged on regular basis through competent international organizations, whether subregional, regional or global where appropriate, and with participation by all states concerned, including States whose nationals are allowed to fish in the exclusive economic zone.

Utilization of the Living Resources

- The coastal state shall promote the objective of optimum utilization of the living resources in the exclusive economic zone..........
- The coastal State shall determine its capacity to harvest the living resources of the exclusive economic zone. Where the coastal State does not have the capacity to harvest the entire allowable catch, it shall.......... give other States access to the surplus of the allowable catch..........
- In giving access to other states..........the coastal State shall take into account all relevant factors including..........the economy of the coastal State concerned and its other national interests..........the requirements of developing States..........and the need to minimize economic dislocation in States whose nationals have habitually fished in the zone or which have made substantial efforts in research and identification of the stocks.
- Nationals of other States fishing in the exclusive economic zone shall comply with the conservation measures..........and regulations of the coastal State. These..........may relate..........to the following:
- Licensing of fishermen, fishing vessels, and equipment including payment of fees..........
- species which may be caught..........Quotas of catch during any period
- Regulating the seasons and areas of fishing, the types, sizes and amount of gear and the numbers, sizes and types of fishing vessels that may be used
- Fixing the age and size of fish..........that may be caught.
- Specifying information required of fishing vessels, including catch and effort statistics and vessel position reports
- Requiring..........fisheries research programmes..........and reporting of associated scientific data
- The placing of observers or trainees on board such vessels
- The landing of all or any part of the catch..........in the coastal State.
- Terms and conditions relating to joint ventures or other cooperative arrangements
- Training of personnel and the transfer of fisheries technology, including fisheries research

- Enforcement procedures
- Stocks occurring within the exclusive economic zones of two or more coastal States or both with exclusive economic zone and in an area beyond and adjacent to it.
- Where the same stock or stocks of associated species occur within the exclusive economic zones of two or more coastal States, these States shall seek..........to agree upon the measures necessary to..........ensure the conservation and development of such stocks..........High migratory species.
- States whose national fish..........for highly migratory species shall cooperate with a view to ensuring conservation..........and optium utilization of such species..........both within and beyond the exclusive economic zone.......... Marine mammal
- Nothing restricts..........the right of a coastal state..........to regulate the exploitation of marine mammals more strictly than provided for in this part. States shall co-operate with a view to anadromous stocks
- States in whose rivers anadromous fish originate shall have the primary interest in and responsibility for such stocks.
- Right of the coastal State over the Continental Shelf
- The natural resources referred to in this part consist of the mineral and other non-living resources of the sea-bed together with living organisms belonging to sedentary species, that is to say, organisms which, at the harvestable stage, either are immobile on or under the sea-bed or are unable to move except in constant physical contact with the sea-bed or the subsoil.
- Pollution from land-based sources.
- States shall..........prevent, reduce, and control pollution from..........rivers, estuaries, pipelines and outfall structures..........
- Pollution by Dumping.

Glossary

Active fishing days – The number of days in a reference period (*e.g.* a calender month) during which fishing activities are "normal" *i.e.* different from zero or not negligible. Often used as an extrapolation factor in catch assessment surveys for the purpose of fishery statistics.

Age class – A group of individual of the same age range in a population. The 0 group are the fish in their first year of life. A fish born in April of a given year, remains in the 0 group until April of the following year. The term usually refers to a year class in long-lived annually breeding species, but shorter units of time are also used, particularly in the tropics.

Age of recruitment – The age when fish are considered to be recruited to the fishery. In stock assessments, this is usually the youngest age group considered in the analyses, typically age 0 or 1.

Age groups – A group of fish at a given age is called age group.

Allowable biological catch – Term used by a management agency which refers to the range of allowable catch for a species or species group. It is set each year by a scientific group created by the management agency. The agency then takes the ABC estimate and sets the annual total allowable catch (TAC).

Annual mortality rate – The ratio between the number of fish which die during a year and the number alive at the beginning of that year.

Area swept – The area of the sea floor over which the gear (usually a trawl, or a dredge) is dragged during its operation. The area is equal to the effective horizontal opening of the gear multiplied by the distance, the gear has covered during the period of time considered (*e.g.* during one hour trawl haul). Combined with information on the fish quantities caught during the considered time period, the area swept allows an estimation of relative or absolute value of the fish density (and biomass) in the area.

Assessment – A judgment made by a scientist or scientific body on the state of a resource, such as a fish stock (*e.g.*, size of the stock, potential yield, whether it is over- or underexploited), usually for the purpose of passing advice to a management authority.

Biological overfishing – Catching such a high proportion of one or all age classes in a fishery as to reduce yields and drive stock biomass, and spawning potential below safe levels. Biological overfishing occurs when fishing levels are higher that those required for extracting the Maximum Sustainable Yield (MSY) of a resource and when recruitment start to decrease statistically.

BMSY – Biomass at MSY. Biomass corresponding to Maximum Sustainable Yield from a production model or from an age-based analysis using a stock recruitment model. Often used as a biological reference point in fisheries management.

Boat day – A measure of fishing effort; *e.g.* 10 vessels in a fishery, each fishing for 50 days would have expended 500 boat-days of effort.

Capture fishery – The sum (or range) of all activities to harvest a given fish resource. It may refer to the location the target resource (*e.g.* fish shrimp species) the technology used (*e.g.* trawl or beach seine), the social characteristics (*e.g.* artisanal, industrial), the purpose (*e.g.* commercial, subsistence, or recreational) as well as the season (*e.g.*) winter.

Catch – To catch: to undertake any activity that results in taking fish (Sensu lato) out of its environment dead or alive. To bringing fish onboard a vessel dead or alive.

Catch-at-age – The estimated number of fish caught, tabulated by fish age and year of capture (and by other strata such as gear or nation). Catch-at-age may be estimated on the basis of catch-at-size, using age-length- keys or cohort.

Catch-at-length – Data on number of fish of each length group in the catch of a fishery, usually obtained by measuring the lengths of fish in representative samples of the catch.

Catch-at-weight – Data on the number of fish of each weight group in the catch of a fishery, usually obtained by measuring the weights of fish in representative samples of the catch.

Catch curve – A graph showing the logarithm of the number of fish taken by fishing at successive ages or sizes.

Catch per unit of effort – CPUE. The quantity of fish caught (in number or in weight) with one standard unit of fishing effort; *e.g.* number of fish taken per 1000 hooks per day or weight of fish, in tons, taken per hour of trawling.

CPUE is often considered an index of fish biomass (or abundance). Sometimes referred to as catch rate. CPUE may be used as a measure of economic efficiency of fishing as well as an index of fish abundance. Also called: catch per effort, fishing success, availability.

Catch rate – Means some times the amount of catch per unit time and sometimes the catch per unit effort.

Catchability coefficient – In general, the extent to which a stock is susceptible to fishing. In fisheries models, the factor (q) relating abundance to stock size (x = q.N) and fishing mortality to fishing effort (F = qf.). Following the later, (q) is a measure of fishing mortality generated on a stock by one unit of effort.

Closed season – The banning of fishing activity (in an area or of an entire fishery) for a few weeks or months, usually to protect juveniles or spawners.

Community – Any naturally occurring group of different organism inhabiting a common environment interacting with each other especially through food relationships and relatively of other groups.

Cohort – In a stock, a group of fish generated during the same spawning season and born during the same time period. A cohort is a batch of fish all of approximately belonging to the same stock.

In cold and temperate areas, where fish are long-lived, a cohort corresponds usually to fish born during the same year (a year class). For instance, the 1987 cohort would refer to fish that are age 0 in 1987, age 1 in 1988, and so on. In the tropics, where fish tend to be short-lived, cohort may refer to shorter time intervals (*e.g.* spring cohort, autumn cohort, monthly cohorts).

Cohort analysis – A retrospective analysis of the catches obtained from a given year class at each age (or length interval) over its life in the fishery. Allows estimation of fishing mortality and abundance at each age as well as recruitment. Involves the use of a simplified VPA algorithm based on an approximation that assumes that, in a given time period, all fishing takes place instantaneously in the middle of the time period.

Conservation – It is defined as the management of the human use of biosphere. So that it may yield the greatest sustainable yield (or) benefit to the present generation. While maintaining it potential to meed the needs and aspirations of future generation as well.

CPUE – Catch per unit of effort. The amount of catch that is taken per unit of fishing effort (*e.g.*, number of fish per longline trawls trus *etc.* hook-moths). Nominal CPUE is often used as a measure of the economic efficiency of a type of gear. Standardized CPUE is normally used as an abundance index for "tuning" assessment models.

Depleted stock – A stock driven by fishing at very low level of abundance compared to historical levels, with dramatically reduced spawning biomass and reproductive capacity. It requires particularly energetic rebuilding strategies and its recovery time will depend on the present condition, the level of protection and the environmental conditions.

Depletion – For renewable resources, the part of the harvest, logging, catch and so forth above the sustainable level of the resource stock; for non-renewable resources, the quantity of resources extracted.

Developed fishery – A "fully" developed fishery is a fishery which, following a period of rapid and steady increase of fishing pressure and catches, has reached its level of maximum average yearly production. It is usually understood that such a fishery is yielding close to it's a maximum sustainable yield.

Developing fishery – A fishery which, after a period of very low activity and landings, increases rapidly and steadily its production through increases in fishing effort, often associated with increased fishing capacity.

Dynamics – The forces which acts on the population (Biotic and Abiotic)

Economic overfishing – This type of occurs when a fishery is generating no economic rent, primarily because an excessive level of fishing effort is applied in the fishery and does not always imply biological overfishing.

Effective fishing effort – (F/q). Fishing effort adjusted, when necessary, so that each increase in the adjusted unit causes a proportional increase in instantaneous rate, of fishing.

***Ex situ* conservation** – Conservation of components of biological diversity out site their natural habitat.

Exclusive economic zone – The area adjacent to a coastal state which encompasses all waters between: (a) the seaward boundary of that state, (b) a line on which each point is 200 nautical miles (370.40 km) from the baseline from which the territorial sea of the coastal state is measured (except when other international boundaries need to be accommodated), and (c) the maritime boundaries agreed between that state and the neighbouring states.

Exploitable biomass – Refers to that portion of a stock's biomass that is available to the fishing gear.

Exploitation rate – The proportion of a population at the beginning of a given time period that is caught during that time period (usually expressed on a yearly basis).

Fertility – Number of eggs actually fertilized.

Fecundity – Number of eggs produced by a female fish.

Fish biomass (Fish abundance/Population size) – It can be expressed either as the number of fish (abundance) (or) the total weight of fish (Biomass (or) size) in a particular area of a given time.

Fish stock – The living resources in the community or population from which catches are taken in a fishery. Use of the term fish stock usually implies that the particular population is more or less isolated from other stocks of the same species and hence self-sustaining. In a particular fishery, the fish stock may be one or several species of fish but here is also intended to include commercial invertebrates and plants.

Fishable stock – The part of a stock that is available to be fished. The fish must be big enough to be caught and must live in places where fishermen work, to be part of the "fishable stock".

Fishers – A word commonly used to describe any male or female engaged in any aspect of the fishing activity.

Fishery – Generally, an activity leading to harvesting of fish activities in which fish is harvested from the wild using some fishing technology (capture fishery) as well as activities producing fish through aquaculture.

Fishery management – The integrated process of information gathering, analysis, planning, decision-making, allocation of resources and formulation and enforcement of fishery regulations by which the fishery management authority controls the present and future behaviour of interested parties in the fisheries, in order to ensure the continued productivity of the living resources.

Fishery models – Simplified representations of the fishery complex reality. May or may not be a mathematical representation.

Fishing – Any activity, other that scientific research conducted by a scientific research vessel, that involves the catching, taking, or harvesting of fish; or any attempt to do so; or any activity that can reasonably be expected to result in the catching, taking, or harvesting of fish and nay operations at sea in support of it.

Fishing effort – The amount of fishing gear of a specific type used on the fishing grounds over a given unit of time *e.g.* hours trawled per day, number of hooks set per day or number of hauls of a beach seine per day. When two or more kinds of gear are used, the respective efforts must be adjusted to some standard type before being added.

Fishing mortality – A mathematical expression of the part of the total rate of deaths of fish due to fishing. Fishing mortality is often expressed as a rate that indicates the percentage of the population caught in a year.

Fishing right – A right to catch a specified quantity of fish, or proportion of the total allowable fish catch or a right to use a boat (or any other specified fishing equipment) in a manner specified in a management plan or in the fishery regulations.

FMSY – The fishing mortality rate which, if applied constantly, would result in Maximum Sustainable Yield (MSY). Used as a biological reference point, FMSY is the implicit fishing mortality target of many regional and national fishery management authorities and organizations. FMSY can be estimate din two ways: (1) from simple production models (2) form age-structured model that include a stock-recruitment relationship.

Frequency distribution – A graphical, tabular, or mathematical representation of the manner in which the frequencies of a continuous or discrete random variable are distributed over the range of its possible values.

Fully exploited – Term used to qualify a stock which is probably neither being overexploited nor underexploited and is producing, on average, close to its Maximum Sustainable Yield. This situation would correspond to fishing at FMSY (in a classical production model relating yield to effort) or Fmax (in a model relating yield-per-recruit to fishing mortality.

Growth – It is the increase in length and weight with respect to time.

A type of input control used as a management tool whereby the amount and/ or type of fishing gear used by fishers in a particular fishery is restricted by law.

Growth overfishing – Growth overfishing occurs when too many small fish are being harvested, usually because of excessive effort and poor selectivity (*e.g.* too small mesh sizes) and the fish are not given the time to grow to the size at which the maximum yield-per-recruit would be obtained from the stock. A reduction of fishing mortality on juveniles, or their outright protection, would lead to an increase in yield from the fishery.

Individual Transferable Quota – ITQ. A type of quota (a part of a Total Allowable Catch) allocated to individual fishermen or vessel owners and which can be sold to others.

Instantaneous rate of growth (g) – The natural logarithm of the ratio of final weight to initial weight of a fish in a unit of time, usually a year, when applied collectively to all fish of a given age in a stock, the possibility of selective mortality must be considered.

Instantaneous rate of mortality – When fishing and natural morality act concurrently, the natural logarithm of the survival rate (with sign changed) for deaths due to either natural causes (instantaneous Rate of Natural Mortality, M)

Or due to fishing mortality (Instantaneous Rate o Fishing Mortality, F, the Instantaneous Rate of Total Mortality, Z, is the sum of these two rates : Z = F + M.

Isopleth – Contour line joining points corresponding to similar values (often yield-per-recruit value (Y/R)) on a graph showing the changes in Y/R as function of size-at-first-capture and fishing mortality.

Length frequency – A length frequency distribution is an arrangement of recorded lengths (in a total catch, stock or a sample) which indicates the number of individuals encountered in each length interval.

Length-frequency distribution – The number of individual of a catch of catch sample in each length interval. The modal size is the length group with the higher number of individual. Distribution may be uni- or bi-modal but are more generally multi-modal, reflecting multiple age-group.

Licensing: Restriction of the right to fish to those persons or vessels issued with licenses for the purpose.

Management – It is all are of coordinating various factors of production to achieve the set goals of the organization through man, materials and machine.

Maximum economic yield (MEY) – When relating total revenues from fishing to total fishing effort in a surplus production model, the value of the largest positive difference between total revenues from and total costs of fishing (including the coast of labour and capital) with all inputs valued at their opportunity costs.

Maximum sustainable yield: The highest theoretical equilibrium yield that cab be continuously taken (on average) from a stock under existing (average) environmental conditions without affecting significantly the reproduction process. Also referred to sometimes as Potential yield.

Mortality – The rate which the individuals are being removed from the fishery. Mortality due to natural (or) fishing.

Mortality rate – The rate at which the numbers in a population decrease with time due to various causes. Mortality rates are critical parameters in determining the effects of harvesting strategies on stocks, yields, revenues, *etc.* The production of the total stock (in numbers) dying each year is called the "annual mortality rate". To facilitate calculations, scientists express mortality as an exponential rate (called instantaneous rates): Nt/N0 = e-Z = e-(M+F) in which Nt/N0 is the survival rate; M, the natural mortality rate; F, the fishing mortality rate; and Z, the total Mortality rate (of deaths due to predation or disease).

Multiple stocks – A stock is said to be a multiple stock when the difference in population characteristics between stocks becomes apparently significant.

Natality – The rate at which new individuals are added to the population by reproduction.

Natural mortality – M. Deaths of fish from all causes except fishing (e,.g. Ageing, predation, cannibalism, disease and perhaps increasingly pollution). It is often expressed as a rate that indicates the per centage of fish dying in a year; *e.g.* a natural mortality rate of 0.2 implies that approximately 20 per cent of the population will die in a year from causes other that fishing.

Nautical mile – Unit of distance equivalent to 1 minute latitude of the great circle of earth (=1852 meters).

Non-renewable resource – Resource can't be replaced

Non-target species – Species for which the great is not specifically set, although they may have immediate commercial value and be a desirable component of the catch.

Ogive – A cumulative per centage frequency distribution.

Optimum yield - OY. A deliberate melding of biological economic, social, and political values designed to produce the maximum benefit to society from a stock of fish.

Other Definitions (MSY or YS.) – The largest average catch of yield that can continuously be taken from a stock under existing environmental conditions. For species with fluctuating recruitment, the maximum might be obtained by taking fewer fish in some years than in others. Also called: maximum equilibrium catch (MEC); maximum sustained yield; sustainable catch.

Over-capitalization – Where the amount of harvesting capacity in a fishery exceeds the amount needed to harvest the desired amount of fish at least cost.

Over-fished – A stock is considered "overfished" when exploited beyond an explicit limit beyond which its abundance is considered "too low" to ensure safe reproduction. In many fisheries for a the term is used when biomass has been estimated to be below a limit biological reference point that is used as the signpost defining an "overfished condition" this sign post is often taken as being FMSY but the usage of the term may not always be consistent.

Production rate – Proportion of population size/biomass

Population – A group of interbreeding organisms that represents the level of organization at which speciation begins.

Population dynamics – They study of dynamics of such population *i.e.* the rate at which they grow, reproduce and die.

Population model – A component of a stock assessment model, made up of formulations that describe how the population changes from one time period to the next. The type of population model used by ICCAT vary, depending on the species life history and on data availability. Population models can roughly be classified as age/size structured or biomass-based; deterministic or stochastic; density-dependent or density-independent; spatially-structured or spatially aggregated; equilibrium or non-equilibrium.

Productivity – The rate at which it is being produced.

Quota – A share of the Total Allowable Catch (TAC) allocated to an operating unit such as a country, a vessel, a company or an individual fisherman (individual quota) depending on the system of allocation. Quotas may or may not be transferable, inheritable, and tradable. While generally used to allocate total allowable catch, quotas could be used also to allocate fishing effort or biomass.

Raising – A procedure for estimating the total from a sample, by multiplying all the fractions in the sample by a "raising factor" equal to the proportion of the total which the sample represents. For example, the total catch at size for fishery is obtained by raising catch-at-size samples to the magnitude of the total catches, *i.e.* by multiplying the sampled numbers times the ratio of sample weight to total catch weight (or the ratio of sample numbers to total numbers).

Recruit – A young fish entering the exploitable stage of its life cycle. A member of "the youngest age group which is considered to belong to the exploitable stock.

Recruitment – The number of fish added to the exploitable stock, in the fishing area, each year, through a process of growth (*i.e.* the fish grows to a size where it becomes catchable) or migration (*i.e.* the fish moves into the fishing area).

The process by which fish enter the exploitable stock and become susceptible to fishing. The process may be short or take more than one year. It is also definitions addition of new fish to the vulnerable population by growth from among smaller size categories.

Recruitment overfishing – A situation in which the rate of fishing is (or has been) such that annual recruitment to the exploitable stock has become significantly reduce. The situation is characterized by a greatly reduced spawning stock, a decreasing proportion of older fish in the catch, and generally very low recruitment year after year. If prolonged, recruitment overfishing can lead to stock collapse, particularly under unfavourable environmental conditions.

Recruits – The new age group of the population entering of the exploited component of the stock for the first time or young fish growing or otherwise entering that exploitable component.

Relative fecundity – Number of eggs produced per unit weight of a fish.

Resources – It is defined as basic (or) primary material of earth's environment which is neither manufactured nor processed product.

Renewable resources: Resource can be replaced.

Renewable natural resource – Natural resources that, after exploitation, can return to their previous stock levels by natural processes of growth or replenishment. Conditionally renewable resources are those whose exploitation eventually reaches a level beyond which regeneration will become impossible. Such is the case with the clear-cutting of tropical forests.

Responsible fisheries – The concept of Responsible Fisheries "encompasses the sustainable utilisation of fishery resources in harmony with the environment; the use of capture and aquaculture practices which are not harmful to ecosystems, resources and their quality; the incorporation of added value to such products through transformation processes meeting the required sanitary standards; the conduct of commercial practices so as to provide consumers access to good quality products.

Ricker W.E. – Computation and interpretation of biological statistics of fish population. Bulletin of the Fisheries Research Board of Canada, 191; 2-6.

A group of individual in a species occupying a well defined spatial range independent of other stocks of the same species. Random dispersal and directed migrations due to seasonal or reproductive activity can occur. Such a group can be regarded as an entity for management or assessment purposes. Some species form a single stock (*e.g.* southern bluefin tuna) while others are composed of several.

Species diversity – The variety of species in a community, which can be expressed quantitatively in ways which reflect both the total number of species present and the extent to which the system is dominate by a small number of species.

State of stocks: An appreciation of the situation of a stock, usually expressed as: protected, underexploited, intensively exploited, full exploited, over-exploited, depleted, extinct or commercially extinct.

A stochastic model is a model whose is under consideration from the point of view of actual or potential utilization.

State: The part of fish population which is under consideration from the point of view of actual or potential utilization.

Stock assessment – The process of collecting and analysing biological and statistical information to determine the changes in the abundance of fishery stocks in response to fishing, and, to the extent possible, to predict future trends of stock abundance. Stock assessments are based on resource surveys; knowledge of the habitat requirements, life history, and behaviour of the species; the use of environmental indices to determine impacts on stocks; and catch statistics. stock assessments are used as a basis to assess and specify the present and probable future condition of a fishery.

Surplus production – It is the weight of fish that can be removed by fishing without causing change in population size.

Survey design – The overall survey design of a probability survey refers to the definitions and the established methods and procedures concerning all phases needed for conducting stock (*e.g.* albacore tuna in the Pacific Ocean comprises separate Northern and Southern stocks). The impact of fishing on a species cannot be determined without knowledge of this stock structure.

Species – A group of actually (or) potentially interbreeding populations which are reproductively separated from other such groups.

Sampling design – The sampling design of a scientific survey refers to the statistical techniques and methods adopted for selecting a sample and obtaining estimates of the survey variables from the selected sample.

Seasonal closure closed season – The banning of fishing activity (in an area or of an entire fishery) for a few weeks or months, to protect juveniles or spawners.

Selective gear – A gear allowing fishers to capture few (if any) species other than the target species.

Selectivity – Ability to target and capture fish by size and species during harvesting operations, allowing by-catch of juvenile fish and non-target species to escape unharmed. In stock assessment, conventionally expressed as a relationship between retention and size (or age) with no reference to survival after escapement.

Size-at-age – Length or weight at a particular age.

Size-at-first-maturity – Length or weight of the fish when it attains maturity as defined by: (1) the minimal size at which maturity is reached; (2) the size at which 50 per cent of the fish of that size are mature.

Spawning biomass – The total weight of all sexually mature fish in the population.

The Survey – The sample design, the selection and training of personnel, the logistics involved in the management of the field force and the distribution and receipt of survey questionnaires and forms, and the procedures for data collection, processing and analysis.

The amount of biomass or the number of units that can be harvested currently in a fishery without compromising the ability of the population/ecosystem to regenerate itself.

TAC – The TAC is the total catch allowed to be taken from a resource in a specified period (usually a year), as defined in the management plan. The TAC may be allocated to the stakeholders in the form of quotas as specific quantities or proportions.

Temperate waters – Waters in the region of higher (cooler; more poleward) latitudes that tropical latitudes; literally those between the Tropic of Cancer and the Arctic Circle in the northern hemisphere, and the Tropic of Capricorn and the Antarctic Circle in the southern hemisphere.

Total mortality rate (z) – Z. Annual or seasonal. The number of fish which die during a year (or season), divide by the initial number. Also called: actual mortality rate, *coefficient of mortality (Heincke).

Unit stock – A population can be considered to be an unit stock if its' members exhibit sufficiently uniform characteristics in respect of their growth, recruitment, spawning, mortality, morphometric and meristic counts *etc.*, *e.g.* Horse mackerel stocks of east coast and west coast due to different growth rate.

Unregulated fishery – A fishery in which producers (and any other participants) are not subjected to any regulations.

Virgin stock – A stock in its natural condition before anyone has fished it.

Virtual Population Analysis – VPA. An algorithm for computing historical fishing mortality rates and stock sizes by age, based on data on catches, natural mortality, and certain assumptions about mortality for the last year and last age group. A VPA essentially reconstructs the history of each cohort, assuming that the observed catches are known without error.

Year class – The fish spawned or hatched in a given year. In the northern hemisphere, when spawning is in autumn and hatching in spring, the calendar year of the hatch is commonly used to identify the year-class (except usually for salmon). Also called: brood, generation.

Yield – The yield curve is the relationship between the expected yield and the level of fishing mortality of (sometimes) fishing effort.

Yield-per-recruit – The expected lifetime yield per fish recruited in the stock at a specific age. Depends on the exploitation pattern (fishing mortality at age) or fish regime (effort, size at first capture) and natural mortality.

Yield-per-recruit analysis – Analysis of how growth, natural mortality, and fishing interact to determine the best size of animals at which to start fishing them, and the most appropriate level of fishing mortality. The yield-per-recruit models do not consider the possibility of changes in recruitment (and reproductive capacity) due to change in stock size. They also do not dean with environmental impacts.

References

Bagenal, T.B. (ed.), 1974. Ageing of fish. Proceedings of an International symposium on the ageing of fish held at the Univ. of Reading, England 19-20 July 1973, Old Working, surrey, England, Unwin Bros. Ltd.,234 p.

Beverton, R.J.H. and S.J. Holt, 1957. On the dynamics of exploited fish populations. Fish.Invest. Minist. Agric. Fish. Food G.B. (2 Sea Fish.), 19;533 p.

Beverton, R.J.H., and S.J. Holt., 1959. A review of the life spans and mortality rates of fish in nature and their relation to growth and other physiological characteristi pp. 142-180.In: GEW Wolstenholme and M.O. Cannor (eds). CIBA Foundation.Colloquia on agencies.Vol. 5.The life span of animals.: London, Churchill, Vol. 5.

Beverton, R.J.H., and S.J. Holt.1966. Tables of yield functions for fishery assessment, FAO Fish. Tech.Pap. (38) Rev.1:49 p.

Blueweiss, L. *et al.,* 1978. Relationship between body size and some life history parameters, Oecologia. 37:257-72

Boonyubol, M, and V.Hongskul, 1978. Demersal reesources and exploitation in the Gulf of Thailand, 1960-1975. In SCS (1978) Report of the workshop on the demersal resources of the Sunda Shelf, Penang, Malaysia, 31 October to 6 November 1977, Part 2. Manila, South China Sea Fisheries Development Programme, SCS/GEN/77/13;56-70

Cushing, D.H., 1968. Fisheries Biology; a study in population dynamics. Madison, University of Wisconsin Press, 200 p.

Cushing, D., 1973. The detection of fish. London, Pergamon Press, 200 p. (Includes some tropical examples)

Devaraj, M. 1983. Fish Population Dynamics, CIFE, Versova, Bombay. 98 p.

Ford, E., 1933. An account of the herring investigations conducted at Plymouth during the years from 1924 to 1933. J. Mar.Biol. Assoc. U.K., 19: 305-384.

Fry, F.E.J., 1949. Statistics of a lake trout fishery Biometrics, 5:27-67.

Gayanilo Jr., F.C., Spare and D. Pauly, 1994.The FAO-ICLARM Stock Assessment Tools (FiSAT) User's guide. FAO Computerized Information Series (Fisheries) No.8 Rome, FAO, 124 p. and 3 diskettes.

Gulland, J.A., 1969. Manual of methods for fish stock assessment.Part 1.Fish population analysis, FAO Man. Fish. Sci., (4): 154 p.

Gulland, J.A., and S.J.Holt, 1959. Estimation of growth parameters for data at unequal time intervals, J.Cons. CIEM, 25(1): 47-9

Gulland, J.A., 1966. Manual of sampling and statistical methods for fisheries biology, Part 1.Sampling methods. FAO.Man. Fish. Sci., spanish.

Holt, S.J., 1957. A method for determining gear selectivity and its application. ICNAFICFS-FAO Joint Scientific meeting paper No. (5): 1-21

Holt, S.J., 1963. A method for determining gear selectivity and its application.Int. Comm. N.W. Atlant Fish.,Spec. Bull No.5 (106-119).

Mesnil, B., 1989. Computer programs for fish stock assessment. ANACO: Software for the analysis of catch data by age group on IBM Pc and compatibles. FAO Fish.Tech. Pap., (101) Suppl. 3:73 p.

Olsen, S. 1959. Mesh selection in herring gillnets. J. Fish Res. Bd. Can. 16(3): 339-349.

Pauly, D., 1981. Tropical stock assessment package for programmable calculators and micro-computers. ICLARM Newsl., 4(3): 10-13.

Pauly, D., 1983. Some simple methods for the assessment of tropical fish stocks. FAO Fish. Tech. Pap., (234): 52p.

Pauly, D., 1984. Fish population dynamics in tropical waters; A manual for use with programmable calculators. ICLARM Stud. Rev., (8): 325p.

Rikhter; V.A. and V.N. Efanov, 1976. On One of the approaches to estimation of natural mortality of fish populations. ICNAF Res. Doc. 76/VI/.18: 12 p.

Robson, D.S. and D.G. Chapman, 1961. Catch curves and mortality rates. Tram Am. Fish Soc. 90(2): 181-189.

Rothschid., 1978. Fishing effort in fish population dynamics, John Wiley, New York.

Schaefer,M., 1954. Some aspects of the dynamics of populations important to the management of the commercial marine fisheries, Bull. I-ATTC, 1 (12): 27-55

Snedecor, G.W. and Cochran., W.G., 1967. Statistical methods. Sixth Edition. Oxford & IBH Publishing Co., New Delhi. 593 pp.

Sparre, P., 1987. Computer programs for fish stock assessment. Length-based fish stock assessment for Apple II computers, FAO Fish. Tech. Pap., (101) Suppl. 2:218 p.

Sparre, P., and S.C. Venema., 1998. Introduction to tropical fish stock assessment. Part I Manual. FAO Fisheries Technical Paper No.306, I, Rev. 2.407 p.

Ziegler, B., 1979.Growth and mortality rates of some fishes of Manila Bay, Philippines, as estimated from the analysis of length-frequencies. Thesis, Kiel University, 115 p.

Venkataramani, K. and Sundramoorthy, B., 2000. Practical manual on Fish population dynamics and stock assessment.

www.ingramcontent.com/pod-product-compliance
Ingram Content Group UK Ltd.
Pitfield, Milton Keynes, MK11 3LW, UK
UKHW021953270726
14060UKWH00002B/497

9 789388 173254